Speculative & Science Fiction

AFRICAN LITERATURE TODAY 39

Guest Editors: Louisa Uchum Egbunike
Chimalum Nwankwo

JAMES CURREY

GUIDELINES FOR SUBMISSIONS

The Editor invites submission of articles on the announced themes of forthcoming issues. Submissions will be acknowledged promptly and decisions communicated within six months of the receipt of the paper. Your name and institutional affiliation (with full postal and email addresses) should appear on a separate sheet, plus a brief biographical profile of not more than six lines. The editor cannot undertake to return materials submitted, and contributors are advised to keep a copy of any material sent. Articles should be submitted in the English language. Articles are reviewed blindly, so do not insert your name, institutional affiliation and contact information on the article itself. Instead, provide such information on a separate page.
Please avoid footnotes or endnotes.
Do not quote directly from the internet without properly citing the source as you would when quoting from a book. Use substantive sources for obtaining your information and depend less on general references.
If your article is accepted for publication, we will require a short (max 50-word) biography for the Notes on Contributors at the front of the volume.

Length: Articles should not exceed 5,000 words.

Format: Articles should be double-spaced, and should use 12pt Times New Roman.

Citations: Limit your sources to the most recent, or the most important books and journals, in English. Cite works in foreign languages only when no English-language books are available. Cite websites only if they are relatively permanent and if they add important information unavailable elsewhere: give both a publication date, and an access date.

Style: UK or US spellings are acceptable, but must be used consistently. Direct quotations should retain the spellings used in the original source. Check the accuracy of citations and always give the author's surname and page number in the text, and a full reference in the Works Cited list at the end of the article. Use single inverted commas throughout except for quotes within quotes which are double. Avoid subtitles or subsection headings within the text.
Italicize titles of books, plays and journals. Put in single quotes titles of short stories, journal articles and titles of poems.
Please follow the ALT Guideline for Submissions and Style Guide: https://boydellandbrewer.com/james-currey-current-authors/.

Copyright: It is the responsibility of contributors to clear permissions for both text and any illustrations used, which should be b/w only. Please note, however, that ALT does not republish chapters or articles that were previously published elsewhere.

All articles should be sent to the editor, Ernest N. Emenyonu, as an email attachment (Word): eernest@umich.edu,
African Literature Today,
Department of Africana Studies,
University of Michigan-Flint,
303 East Kearsley Street,
Flint MI 48502, USA.
Fax: 001-810-766-6719

Books for review to be sent to the Reviews Editor:
Obi Nwakanma, University of Central Florida,
English Department,
Colburn Hall,
12790 Aquarius Agora Drive,
Orlando, FL 32816, USA
Obi.Nwakanma@ucf.edu

AFRICAN LITERATURE TODAY

ALT 1–14 BACK IN PRINT. See www.jamescurrey.com to order copies

ALT 1, 2, 3, and 4 Omnibus Edition
ALT 5 The Novel in Africa
ALT 6 Poetry in Africa
ALT 7 Focus on Criticism
ALT 8 Drama in Africa
ALT 9 Africa, America & the Caribbean
ALT 10 Retrospect & Prospect
ALT 11 Myth & History
ALT 12 New Writing, New Approaches
ALT 13 Recent Trends in the Novel
ALT 14 Insiders & Outsiders

Backlist titles available in the US and Canada from Africa World Press and in the rest of the world from James Currey, an imprint of Boydell & Brewer

ALT 15 Women in African Literature Today
ALT 16 Oral & Written Poetry in African Literature Today
ALT 17 The Question of Language in African Literature Today
ALT 18 Orature in African Literature Today
ALT 19 Critical Theory & African Literature Today
ALT 20 New Trends & Generations in African Literature
ALT 21 Childhood in African Literature
ALT 22 Exile & African Literature
ALT 23 South & Southern African Literature
ALT 24 New Women's Writing in African Literature
ALT 25 New Directions in African Literature

Recent and forthcoming titles*

Nigeria edition (ALT 24–33): HEBN Publishers Plc
African Literature Today continues to be published as an annual volume by James Currey (an imprint Boydell & Brewer since 2008).
North and South American distribution:
Boydell & Brewer Inc., 68 Mount Hope Avenue, Rochester, NY 14620-2731, US
UK and International distribution:
Boydell & Brewer Ltd., PO Box 9, Woodbridge IP12 3DF, GB.

ALT 26 War in African Literature Today
ALT 27 New Novels in African Literature Today
ALT 28 Film in African Literature Today
ALT 29 Teaching African Literature Today
ALT 30 Reflections & Retrospectives in African Literature Today
ALT 31 Writing Africa in the Short Story
ALT 32 Politics & Social Justice
ALT 33 Children's Literature & Story-telling
ALT 34 Diaspora & Returns in Fiction
ALT 35 Focus on Egypt
ALT 36 Queer Theory in Film & Fiction
ALT 37
ALT 38 Environmental Transformations

Speculative & Science Fiction

AFRICAN LITERATURE TODAY 39

James Currey
is an imprint of
Boydell & Brewer Ltd
PO Box 9, Woodbridge
Suffolk IP12 3DF (GB)
www.jamescurrey.com
and of
Boydell & Brewer Inc.
668 Mt Hope Avenue
Rochester, NY 14620–2731 (US)

www.boydellandbrewer.com

First published 2021

British Library Cataloguing in Publication Data
A catalogue record for this book is available from the British Library

ISBN 978-1-84701-285-2 (James Currey hardback)

Contents

*** STOP PRESS/TRIBUTE TO NAWAL EL SAADAWI (1931–2021)
Prolific and versatile ultra-socialist feminist Egyptian writer NAWAL EL SAADAWI, popularly described as the spokesperson of women in the Arab World and a relentless advocate for human rights and social justice for women the world over, died while we were going to press with this volume. We will carry a tribute to her and her legacy in the next issue, *ALT 40*.

Notes on Contributors

Austine Amanze Akpuda is an Associate Professor in the Department of English Language and Literature, Abia State University, Nigeria. He has written meta-critical essays on Ngũgĩ wa Thiong'o, Nawal El Saadawi, and Sam Ukala, among others. He also edited *Reconstructing the Canon: Festschrift in Honour of Proffessor Charles E. Nnolim* (2001) and co-edited, with Greg Mbajiorgu, *50 Years of Solo Performing Art in Nigerian Theatre, 1966-2016* (2018), published by Kraft Books.

Gabriel Bámgbóṣé is a Nigerian poet, critic and translator, and a PhD Candidate in Comparative Literature at Rutgers University-New Brunswick. He has taught English and Yorùbá at Tai Solarin University of Education, Nigeria, and New York University, New York. Bámgbóṣé's writing has appeared in *CLCWeb: Comparative Literature and Culture* and *Contemporary Humanities*, among others. He is the author of the poetry collection *Something Happened After the Rain*.

Ezeiyoke Chukwunonso is currently a PhD student at Manchester Metropolitan University, where his research focuses on the impacts of postcolonial protest literature on the evolution of Africa Speculative Fiction. His collection of short stories, *Haunted Grave and Other Stories*, was published by Parallel Universe Publications.

Iquo Diana Abasi writes prose and scripts for radio and screen. Her book, *Èfó Rírò and Other Stories* was released in 2020, and her first collection of poems, *Symphony of Becoming*, was shortlisted for the Nigeria LNG prize for literature, the ANA poetry prize and the Wole Soyinka Prize for Literature. Iquo resides in Lagos, where she edits Omenana.com, a speculative fiction magazine.

Jeffrey G. Dodd is Assistant Professor of English Department at Gonzaga University in Spokane, Washington.

Arthur Onipede Hollist (aka Pede Hollist) is an award-winning fiction writer and African literature professor at The University of Tampa. His interests cover the literary and critical expressions of the African continent and diaspora. He has published one novel, multiple short stories, and critical articles in peer-reviewed periodicals. He was a Fulbright Scholar to Sierra Leone from 2017–2018.

Elizabeth L.A. Kamara holds an Honours degree in English Language and Literature and a Master of Arts degree, both from the University of Sierra Leone. She is the Head of the English Unit in the Department of Language Studies at Fourah Bay College, the University of Sierra Leone. She is the author of two collections of poems and is married with two lovely sons.

Eshitika L. Lutomia is a passionate writer whose works expose Kenya to the rest of the world. His story 'Murmurs of the Spring' was published in issue 2 of *Kikwetu* journal in 2016. Eshitika was born in Butere, Kenya and attained his BA (Creative Writing and History) in Nairobi, where he now lives.

Edgar Fred Nabutanyi holds a PhD in English from the Department of English, Stellenbosch University South Africa and he is a Senior Lecturer and Head of the Department of Literature, Makerere University. His research interest is in how contemporary Ugandan fiction – novels, verisimilitude, and sci-fi short fiction – empowers vulnerable minorities to disclose their agency and subjectivity.

Kayode Odumboni is a PhD student in the Department of English at the Ohio State University.

Reginald Chiedu (R.C.) Ofodile has had a varied career as a lawyer, actor and writer. He has worked internationally on stage, screen and radio, and has published novels, plays, poetry collections, and a short story collection. His awards include the Warehouse International Playwriting, the BBC African Performance and prizes from the Association of Nigerian Authors for his short story collection *From Sin to Splendour* (2016), and for his play, *Magnetism* (2017).

Ifeoma Okoye is a retired university lecturer in English. She attended University of Nigeria, Nsukka, and Aston University, Birmingham, UK. Among her many awards, Ifeoma's novel *Men Without Ears* (Longman 1984) won Best Fiction of the Year from the Association of Nigerian Authors. Her children's books have been translated into French and six African languages. Her husband, Mokwugo Okoye, a writer and frontline nationalist died on 21 September 1998.

Tijani Abdullahi Olaniyi is a poet and a student of politics at the University of Ibadan. Raised in the South West of Nigeria, he is a vibrant and energetic young writer. He is an ambassador of humanity and an advocate of societal reformations. His writings cut across love, African Society, Social Emancipation, and more.

James Orao teaches Intercultural German Studies at the University of Nairobi (Kenya). His PhD was on *Narrated Interculturality – Migration and Identity in Contemporary German and English Language Literature*. His teaching and research focus includes contemporary German-language literature, intercultural literary studies and contemporary literatures that construct subversion and subversive narratives.

Clara Ijeoma Osuji holds a PhD in Literature from the University of Lagos, Nigeria. She teaches Literature in the Department of English, Augustine University, Nigeria. Some of her research efforts have been published both locally and internationally, in journals and edited volumes. Her ongoing scholarly projects focus on aging masculinities.

Niyi Osundare is a Nigerian poet, playwright, essayist, scholar, and winner of numerous literary prizes and awards. He is a Laureate of the Nigerian National Order of Merit (NNOM), Nigeria's highest award for distinguished creative and intellectual achievement. He is currently Emeritus Distinguished Professor of English, University of New Orleans, USA, and Visiting Professor, University of Ibadan, Nigeria.

Janelle Rodriques is an assistant professor of English at Auburn University, Alabama, where she specializes in Caribbean and Black Atlantic Literature. Her monograph, *Narratives of Obeah in West Indian Literature* (2019), is published with Routledge.

Kadija George Sesay, FRSA, is the co-founder of Mboka Festival of Arts Culture and Sport in The Gambia. Her poetry collection is *Irki*. She published *SABLE LitMag* for 15 years until 2015 and is currently the Publications Manager for Inscribe, a writer development programme at Peepal Tree Press which will publish the first anthology of Black British speculative fiction in 2022. She is a PhD scholarship student at Brighton University researching Black British Publishers and Pan-Africanism.

Éric Essono Tsimi is Assistant Professor of Francophone Studies at Baruch College, The City University of New York, specializing in literary theory, Subsaharan literary and cultural studies, and Black France. *Vous autres, civilisations, savez maintenant que vous êtes mortelles* (2021), his most recent book is specifically dedicated to counter-utopia as a French literary genre.

Aisha Umar is a novelist, poet, literary critic and PhD student with the University of Ilorin. She lectures with the Department of European Languages, Federal University Birnin Kebbi, Nigeria. She specializes in gender and teaches African Literature, Ecocriticism and Creative Writing. Her first novel '*Yar Fari,* is published by Kraft Books.

Kufre Usanga is a graduate student in the Department of English & Film Studies, University of Alberta. Her area of research is Indigenous literatures and Petrocultures. Her interests include African feminisms and oral literatures. Her book reviews have been published in various online literary magazines as well as in *African Literature Today* and *ARIEL*.

Introduction

Speculative & Science Fiction: What is *Past & Present* … & What is *Future?*

LOUISA UCHUM EGBUNIKE & CHIMALUM NWANKWO

For this fascinating edition of *African Literature Today*, it is useful to suggest this guiding broad picture as heuristic genre marking. The speculative is trapped in the realm of suggestion. It is a half-birth in terms of a tactile reality. Science fiction on the other hand, through systematic practice, appears to make sure that what is suggested as imagined enjoys strong tactile and kinesthetic probability and full birth. That was how the Greek word *techne/tekne* anticipated the system from where the word technology became what we know today. African speculative fiction may tell how the tortoise or spider travelled to heaven to encounter God. Science fiction, typified by works by writers like Jules Verne and H.G. Wells, stakes a different trajectory and terminus. Their creative mantra rode on, as the saying goes, what one man can imagine, another man can create. In other words, you do not just imagine or speculate without the possibility of creating and actualizing. Clearly not so for speculative fiction which only projects open-ended possibilities. Unless we get this distinction clear, the missions of the two enterprises and their different goals will keep causing needless debate and confusion with the concomitant comparison of scales of importance or the familiar binary of inferiority/superiority. It is ostensible that those writing about the tortoise going to see God did not intend that mission to be actualized. Those whose writings anticipated submarine warfare and journeys to the moon and elevators and so forth before rockets and submarines and elevators became real believed that technology could make them be or happen.

What is speculative fiction and which is science fiction? What is past and present? And what is future? We may proffer a-racial or a-political answers. In the end, the answers under careful scrutiny with appropriate trenchant scalpels are likely to be found wanting with all the various labels having no easy escape from the tags of spurious or questionable because of the frame. Here, for instance, is our distinction between the speculative and science fiction coming out of our ruminations so far. What is obvious from a wholesome African way is that the African now is both speculative and scientific very much like the African modernity or postmodernity. The much talked about epistemological disruptions are academic. Ontological disruptions? Yes, maybe, but read together, the articles by Dodd and Odumboni, Chukwunonso, and Tsimi answer some or most of the questions associated with those and other suggested disruptions.

During a series of interviews conducted in 2017 for a BBC Radio 3 feature on 'Afrofuturism', focusing on whether the term was expansive enough to encompass African cultural production, a number of African writers and filmmakers expressed their views on the subject. While rooted in African-American artistic expression, the term 'Afrofuturism' had gained credence in its application to African speculative works. Coined by cultural critic Mark Dery, Afrofuturist works are described as '[s]peculative fiction that treats African-American themes and addresses African-American concerns in the context of twentieth-century technoculture – and, more generally, African-American signification that appropriates images of technology and a prosthetically enhanced future' ('Black to the Future': 180). The application of the term to works emerging from Africa and its recent diaspora, and its retrospective employment to earlier works of African literature, signalled an elasticity to the term, as it was stretched to incorporate broader literary traditions and experience. The initiation of this conversation recognized the expansion in films and literature being created by Africans on the continent and its recent diaspora, which sat somewhere within the continuum of the speculative arts. By embracing the term 'Afrofuturism', there was an entangling of literary genealogies and traditions, which although historically tied, were at the same time distinct from each other. Two parallel conversations emerged; one that centred on solidarity and collectivity, recognizing the historical processes of erasure and marginalization, and a second conversation that recognized the distinct cultural heritage and literary forebears of African writers and their African-American counterparts.

For filmmaker Wanuri Kahiu, the initial resistance she describes having in the wake of her work being classified as 'Afrofuturist' was not centred on whether the term was expansive enough to incorporate African forms of creative expression, but rather the generic designation of her work, based on cultural categories or geographical locations:

> I only knew about Afrofuturism years after *Pumzi* came out. I wouldn't have termed my work as Afrofuturism, nor would I have termed *Pumzi* as Afrofuturism. I would have perhaps described it as science fiction. The idea of boxing it in a location or an ethnicity or a culture was a bit disturbing at the beginning, because I was just making a film that was set in the future. It had nothing to do with me being an African when I started creating. The more I explored the idea of what Afrofuturism was, I began to embrace it a little more, only because there was something that Mark Dery, who was one of the people who coined the phrase Afrofuturism said. And he said that because Black people have had themselves written out of their own history, they now very intentionally write themselves into their own future. And I thought that was a really interesting take on Afrofuturism, and I decided to adopt that kind of ethos, of being somebody who actively writes ourselves into the future. ('Sunday Feature')

The discomfort that accompanied Kahiu's survey of critical responses to her work echoes some of the longstanding tensions that exist within African literary spaces. The tendency to categorize creative works along national, regional or ethnic lines is perceived by some writers as an imposition at best or a form of ghettoization at worst. Dambudzo Marechera famously resisted such forms of categorization, declaring that '[i]f you are a writer for a specific nation or a specific race, then fuck you' (Veit-Wild and Schader, *Dambudzo Marechera*: 3). The relationship between writer and critic has at times been fraught, as critics attempt to locate writers and their works within critical frames which may not align with the writer's own conception of themselves, or their works. Kahiu's eventual embrace of the term Afrofuturism demonstrates its shifting parameters, but also signals a form of generic solidarity, as Kahiu aligns her work with the political objectives of Afrofuturism, in acknowledging the historical degradation of African peoples, and in making a deliberate effort to produce narratives of the future which centre Black people. In the exploration of critical frameworks through which to engage African speculative fiction, this volume of *African Literature Today* is involved in an intellectual project that is located within the broader and longstanding conversation on how to engage with African writing, and which paradigms expedite the process.

The presence of writers in the course of defining and utilizing terminology is nothing new, and has in fact been central to the development of contemporary African literary criticism. In the 1962 'Conference of African Writers of English Expression' at Makerere University, one of the central discussions that emerged was what has come to be known as the 'language debate'. This discussion centred on the language of African literature; as the question was posed, can works contributing to the African literary canon be written in a European language? The conversation spilled over onto the pages of *Transition Magazine,* of which the editor Rajat Neogy, had been a delegate at the conference. Obi Wali's commentary 'The Dead End of African Literature' made a strong case for writing in indigenous languages, suggesting that writing in European languages contributed to European canons at the expense of the development of Afrophone literature. Gabriel Okara's article 'African Speech … English Words' centred on the process of translation and transliteration from his native Ijaw into English, which constituted part of his writing process. The language debate extended far beyond the confines of the conference and *Transition Magazine*, featuring prominently in subsequent writings of Ngũgĩ wa Thiong'o and Chinua Achebe, both of whom had attended the conference, and occupied opposing sides of this discussion.

The conference, described by Ngũgĩ in 'A Kenyan at the Conference'[1] as 'a landmark in the cultural reawakening of our continent' (7), revealed two things that may prove useful in mapping contemporary discussions of identifying useful definitions for African speculative works. These points relate to the relationship between African and African-American literature, and the absence of Amos Tutuola at the conference. In spite of the conference's title, conversations on West Indian literature were scheduled, bolstered by the presence of Jamaican playwright, Barry Reckford. As John Nagenda relays in his notes on the conference, '"The American Negro Scene" was introduced by two eminent Negro writers, Langston Hughes and Saunders Redding' (8). Langston Hughes would go on to be declared the conference's guest of honour, with Es'kia Mphahlele delivering 'a five-minute tribute that seemed to stun and embarrass Langston' (Rampersad, *The Life of Langston Hughes*: 354), after which Redding read a selection of his poems. What these episodes reveal is that while this was a curated space for conversations on the emergence of modern African literature, this conversation was placed in tandem with African diasporic cultural production. Debates around the emergence of African literature were in dialogue with African-American and West Indian writing, in recognition of the connectivity of these histories and traditions. These connections were particularly pertinent given the socio-political concerns of the time on both sides of the Atlantic, chiefly the civil rights movement in the United States and the anti-colonial and independence movements in Africa and the Caribbean. The second notable observation is the absence of Amos

Tutuola, the first Nigerian to publish a novel, and arguably Africa's most prominent writer of speculative works at the time. Tutuola's *The Palm-wine Drinkard*, published in 1952, introduced the wider reading public to folkloric tales rooted in the Yorùbá oral literary tradition. Obi Wali notes that '[n]ot only was Tutuola, who undoubtedly is one of the most significant writers in Africa today, not present in the conference, but there was a careful exclusion of his works in the discussions of the conference' ('The Dead End of African Literature?': 13). When delivering a talk in Ibadan in 1977, Achebe recalls how a Nigerian student in America, on hearing that he taught *The Palm-Wine Drinkard*, made it clear that this was 'not the kind of thing a patriotic Nigerian should be exporting to America', viewing Tutuola's work as 'childish and crude' (*Hopes and Impediments*: 100).

It lends credence to ideas of 'serious writing' at the inception of this literary tradition; the language of Tutuola's writing, shaped by his standard six primary education, contributed to its hostile reception among some sections of Africa's middle class. Tutuola's deep rootedness in folklore set his writing apart from the 'realist' works which were emerging as the standard for contemporary African writing. As Mũkoma wa Ngũgĩ notes, 'Tutuola's style and English usage were seen not as setting or contributing to a literary tradition, but rather as a one-off style that had no future' (*The Rise of the African Novel*: 19). Thus, speculative writing largely existed at the margins of African writing (with a notable exception in Ben Okri's writing, although much later). Mũkoma notes that '[w]ith Achebe joining AWS as editorial advisor in 1962, the series would go on to set the African literary tradition on the path of realist novels' (19).

Almost sixty years later, writers are once again congregating to discuss literary traditions, genre and literary genealogies, with a rise in speculative works being produced by Africans on the continent and in the recent diaspora. Many of these conversations, however, are taking place in digital spaces through publications such as *Brittle Paper*, *Omenana*, or blog posts by writers themselves. Historically, the academy's primary focus on realist literature has meant that the treatment of speculative works has received less attention, with a tendency to focus on particular texts or writers, rather than an in-depth study on speculative writing in its various forms.[2] There has been a tendency to discuss African speculative works within paradigms that are rooted in distinct literary traditions, be it the largely African-American tradition of Afrofuturism, or 'magical realism' which is rooted in fantastical works from South and Central

America. Over the last decade, there have been a number of journal special issues engaging African speculative fiction within the specific generic frames of African Science Fiction, opening up new conversations in the field, notably *Paradoxa*'s special issue on African SF (2014) edited by Mark Bould, *Cambridge Journal of Postcolonial Literary Inquiry*'s special issue 'African Science Fiction' (2016) edited by Moradewun Adejunmobi and *Vector*'s special issue 'African and Afrodiasporic SF' (2019).

In recent years there has been greater scrutiny placed on the term 'Afrofuturism' in relation to African speculative works, and this has coincided with Nnedi Okorafor's coining of 'Africanfuturism' (as distinct from 'African Futurism') and 'Africanjujuism'. Okorafor, who at a point accepted the term 'Afrofuturism' as a descriptor of her works, explains the shift in her thinking, noting that as the term did not fully capture her creative and political project, it led to the misinterpretation of her works. Okorafor also cites wanting to exert some agency over how her work is discussed. She explains that 'Africanfuturism is a sub-category of science fiction. Africanjujuism is a subcategory of fantasy that respectfully acknowledges the seamless blend of true existing African spiritualities and cosmologies with the imaginative'. Okorafor notes the similarities with Afrofuturism, and describes the distinction in the following way: 'The difference is that Africanfuturism is specifically and more directly rooted in African culture, history, mythology and point-of-view as it then branches into the Black Diaspora, and it does not privilege or center the West'. (Okorafor 'Africanfuturism Defined') The generic categories set out by Okorafor, one of Africa's most prolific and successful contemporary writers of speculative fiction, has seen largely a positive response to her suggestions. The framework of 'Africanfuturism' was adopted by Wole Talabi in his curation and editing of an anthology of short stories. In his introduction, Talabi notes that

> Africanfuturist stories going as far back as the history of the genre can (and should) now be clearly seen and read through a lens that centres them and their viewpoints, encouraging readers around the world to actively engage with African traditions of thought, of science, of philosophy, of history, of dreams, of being. We believe there is value in this focus, in this clarity. (Talabi 'Introduction')

Through his adoption of the term Africanfuturist, Talabi is contributing to the shaping and expansion of the concept, in that this anthology provides us with examples of Africanfuturist writing beyond Okorafor's writing alone. Moving away from the term 'African Science Fiction' which was adopted by Ivor Hartman for the anthology of stories he edited in 2012, Talabi instead recognizes a particular usefulness in Okorafor's framing:

> Science is not Western, despite what colonial history would have us think … When we think of science, we think of computers, robots, Western medicine, all these kinds of tropes. But no, no. Science is a way of understanding the world that is evidence based. … This is what makes science essentially different from many religions … In the understanding of plants, for example, many of our ancestors in many parts of Africa, they did apply a very evidence based approach to understanding the world around them. They will try different herbs to see what works … then that became the remedy and they would encode this in songs or in the way they taught their apprentices … This is the science based thinking, to come back to science fiction and Africanfuturism specifically. And why I find the term useful is because there are certain science fiction stories, science fiction in the tradition I've just described that respect science and scientific thinking and evidence based thinking. (Talabi, correspondence with the author)

Talabi demonstrates the rootedness of science (and by extension, science fiction) in the African reality, and champions the formulation of terminology which captures the specificity of African experiences. In this neologism, Africa and futurity are permanently bound together, signalling that this has always been part of an African reality.

In this volume, many of the contributors have thought carefully about terms and terminology, offering up multiple critical lenses through which to read a broad range of fiction. Ranging from the application of the broad term 'speculative fiction', to an engagement with the specific sub-genres of Africanfuturism, Africanjujuism, counter-utopian novels and political futurism, the contributors have grappled with these concepts at both the level of theory and praxis. The intention of this special issue is to open up conversations on African speculative fiction, examining critical frameworks and their applications. It features the writings of scholars working in these fields, placed alongside interviews with writers who are at the helm of creating speculative works of fiction. At this juncture in the writing and study of African speculative fiction, questions of literary traditions, generic boundaries and appropriate terminology have informed the discussions that ensue, from calls to revisit canonical texts and re-read them through a speculative lens, to suggestions of expanding the canon to include more speculative works.

Okorafor's break from 'Afrofuturism' is articulated in her TED talk: 'My science fiction has different ancestors – African ones' (Okorafor 'Sci-fi Stories that Imagine a Future Africa'). In this volume, James Orao's paper describes the evolution of Okorafor's thought, remarking on the significance of her shift from 'Afro' to 'African' as a movement 'away from a hyphenated space defined by Blackness solely in relation to whiteness, to a 'purely' African space, both on the continent as well as the diaspora'. For Orao, this break is framed as a liberating one as it moves to an Africa-centred space. Likewise, Janelle Rodriques suggests that 'Africanfuturism is inward looking; it does not concern itself with writing back but instead, writes itself *in*'. Rodriques identifies this movement from the peripheries to the centre, as bound up with the very real experience that

> [h]istory in Africanfuturist narratives does not have to be 'uncovered' or 'made legible' in the same ways; the futures that are imagined are less 'impossible', as they are known to have existed. African futures are rooted in tangible pasts that pre-exist Anglo-European colonisation, and as such do not seek to reverse the 'social death' of slavery.

For Kayode Odumboni, 'Africanfuturism engages speculative fiction with an unambivalent aim of disrupting the epistemological structures upon which the history of colonialism and the present realities of neocolonialism subsist. Therefore, Africanfuturism offers a scathing postcolonial, counterhegemonic critique of global modernity'. Orao, Rodriques, and Odumboni provide thoughtful engagements with Africanfuturism in their respective works, building on the critical foundations laid out by Okorafor. Each of these authors views Africanfuturism as occupying a space of strength which resides in its rootedness in Africa of the past, present and future.

While Africanfuturism and Africanjujuism remain relatively new terms, the malleability and adaptability of these terms are explored by the contributors. Rodriques creates her own neologism through combining these two distinct sub-genres into Africanjujufuturism. This all-encompassing term is able to speak across Africa-centred science fiction and fantasy writing, providing an avenue through which to traverse the specific sub-genres to provide both specific and broader points of reference. Likewise, Jeffrey Dodd's coinage 'Africannearfutures' plays on 'Africanfuturism', in that

> [i]f an Africanfuturist lens allows us to … move beyond 'what could have been' towards 'what is and can/will be', the ambivalences revealed in the relationships between (neo-/post-)colonial subjects and the

> technologies they create provides a starting point for imagining the slippages in those futures.

Dodd continues

> [w]hile such ambivalences and slippages are present in Africanfuturist texts exploring visions of distant futures, they are arguably more salient in those that imagine futures closer to our contemporary time. To highlight this distinction and focus in our analysis, I will use 'Africannearfutures' as a playful adaptation of Okorafor's term Africanfuturism.

This expansion and extension of the language opens up new possibilities for engaging literary texts, and contributes to the broader project of developing a suitable language through which to engage African speculative writing.

Clara Osuji's paper in this collection returns to the African literary canon, as she proposes a re-reading of Chinua Achebe's *Things Fall Apart* and *Arrow of God* through a speculative lens. Osuji notes that '[w]hile Achebe's classic, *Things Fall Apart*, and his third novel, *Arrow of God,* have been subjected to various intellectually convincing interpretations, this chapter attempts to further subject the texts to a new demand by linking them to African speculative fiction'. This return to canonical works, primarily discussed within the realist tradition, troubles some of the generic boundaries and borders that have been enforced, while arguing that although a work may not be speculative in its entirety, many African works have 'speculative threads' which run through them. Edgar Nabutanyi also returns to the canon in his contribution, and considers how the presence of recent speculative writing in Uganda can and should disrupt the conventions of Uganda's literary canon. Both Osuji and Nabutanyi make strong cases for expansions: whether it is in how we approach the reading of African literary classics, or what we regard as canonical works.

Returning to the works of Kojo Laing, Ezeiyoke Chukwunonso suggests that, in misreading Laing's works as magical realism, scholars have unfairly classified his writing as 'difficult' to engage. Chukwunonso proposes reading Laing through a 'political futurism' lens, which

> can either be dystopic or utopic. It positions itself against the scientism and progressivism inherent in some 'technocultures' that are underpinned by capitalist ideology with a naïve faith that progressiveness or superiority in technology will end inequality and environmental problems. 'Political futurism' looks at societal webs or what Arne Johan Vetlesen calls 'exchange and mutual dependency'.

Chukwunonso argues that the tendency to read Kojo Laing's writing as magical realism results in an overlooking of the politics of Ujamaa embedded in his writing. Rather than engage in the politics of 'writing back' which Chukwunonso suggests keeps the writer trapped in a dialectical exchange with the colonizer, He emphasizes the political and economic dimension of Laing's approach to decolonization. Moving beyond the Anglophone world, Éric Essono Tsimi's chapter introduces the reader to 'counter-utopias' in Francophone African writing. Tsimi explains that '[t]he counter-utopia is an objective variant of utopia: the difference lies in its worthiness being that it is not the absolute best but the relative best that one imagines'. The counter-utopia in Francophone writing is 'distinct from Anglophone dystopias' in that they 'talk of a future time that is lost in the present times' and speak about 'a present that escapes us through a possible future'. The shifting temporal dimensions are matched by a physical space that also is in a state of flux, reminding us that conversations of futurity and temporality inevitably invoke discussions of spatiality. Finally, note carefully the peculiar immortality implied in Kojo Laing's deathless story, *Woman of the Aeroplanes*, addressed quite insightfully by Chukwunonso, the great story of resolution through the acceptance of needless levels and levelling, of simultaneity and contemporaneity within one reality or plane – Laing's paradise of Tukwan and Levensvale. So indeed, whether it is about humanity or deities, about the natural or the supernatural, about here or hereafter, the planet and its furthest reaches, *speculated*, is imbued with a *holistique*, with something ecumenical in its albumen, something truly *immortal.*

The eight articles in this volume of *African Literature Today*, *Speculative & Science Fiction*, make a case for the imaginative possibilities enabled through speculative writing. The contributors demonstrate how African writers have engaged the speculative to imagine beyond the confines of what is 'real' while rooting their imaginings in the cultural heritage of African peoples. This edition of ALT has to be read and studied carefully in order for one to appreciate the mission of these contributions, their successes in drawing attention to critical aspects of the African imagination and its cultural productions, and the aspects of all that which in terms of clarity or clarifications should still stand as works in progress, but definitely not progress towards any superior genre but towards a perfection of craft. By and large, a quest is on. That quest is for an *African holistique*.

NOTES

1 At this point in his career, Ngũgĩ wa Thiong'o was publishing under the name 'James Ngũgĩ', and so this is the name that appears in this editorial's 'WORKS CITED' section.

2 A notable exception is Brenda Cooper's *Magical Realism in West African Fiction: Seeing with a Third Eye* (London: Routledge, 1998).

WORKS CITED

Achebe, Chinua. *Hopes and Impediments: Selected Essays*. New York: Doubleday, 1989.

Adejunmobi, Moradewun (ed.) 'African Science Fiction'. *Cambridge Journal of Postcolonial Literary Inquiry*, special issue, Vol. 3, No. 3, 2016.

Bould, Mark (ed.). 'Africa SF'. *Paradoxa*, special issue, No. 25, 2014.

Clarke, Michelle Louise (ed.). 'African and Afrodiasporic SF'. *Vector*, special issue, No. 289, 2019.

Dery, Mark. 'Black to the Future: Interviews with Samuel R. Delany, Greg Tate, and Tricia Rose'. In *Flame Wars: The Discourse of Cyberculture*, Mark Dery (ed.). Durham NC: Duke University Press, 1994: 179–222.

Kahiu, Wanuri. Interview in 'Sunday Feature: Louisa Egbunike and Sean Williams', *BBC Sounds*, 2017. www.bbc.co.uk/sounds/play/b09bx511 (accessed 26 February 2021).

Nagenda, John. 'Conference Notebook'. *Transition* No. 5, 1962: 8–9.

Ngũgĩ, James. 'A Kenyan at the Conference', *Transition* No. 5, 1962: 7.

Okara, Gabriel. 'African Speech … English Words'. *Transition* No. 10, 1963: 15–16.

Okorafor, Nnedi. 'Africanfuturism Defined'. Nnedi's Wahala Zone Blog, 19 October 2019. http://nnedi.blogspot.com/2019/10/africanfuturism-defined.html (accessed 1st March 2021).

—— 'Sci-fi Stories that Imagine a Future Africa'. *TED*, August 2017. www.ted.com/talks/nnedi_okorafor_sci_fi_stories_that_imagine_a_future_africa (accessed 24 February 2021).

Rampersad, Arnold. *The Life of Langston Hughes, Volume II: 1941–1967: I Dream a World*. Oxford: Oxford University Press, 2002.

Talabi, Wole. Correspondence with Louisa Uchum Egbunike 8 March 2021 (unpublished).

—— 'Introduction'. In *Africanfuturism: An Anthology*, Wole Talabi (ed.), 2020. http://brittlepaper.com/wp-content/uploads/2020/10/Africanfuturism-An-Anthology-edited-by-Wole-Talabi.pdf (accessed 3 March 2021).

Veit-Wild, Flora and Ernest Schade (eds). *Dambudzo Marechera: 1952–1987*. Harare: Baobab Books, 1988.

Wali, Obiajunwa. 'The Dead End of African Literature?' *Transition* No. 10, 1963: 13–15.
wa Ngũgĩ, Mũkoma. *The Rise of the African Novel*. Ann Arbor: University of Michigan Press. 2018.

Articles

'Being very human in one of the most inhuman cities in the world'

Lagos as a Site of Africanfuturist Invasion in *Lagoon & Godhunter*[1]

JANELLE RODRIQUES

In 1982, Salman Rushdie's report on an Asian writers' conference was titled, by the editor of *The Times* of London, 'The Empire Writes Back with a Vengeance'. (8) This rather sensationalized headline has become paradigmatic of 'postcolonial' – read non-Anglo-Saxon – writing and, to this day, experiments in English-language writing by writers of colour are still considered 'rebellious' by the literary establishment. This paradigm imagines aggression where it might not exist, and insurrection where there is merely expression. Yet, many writers of African futures today have abandoned this particular posture of 'desecration' – if ever they took it up – and turned their foci inward. The futures they imagine have broken away from the toxic embrace of neoliberal – neocolonial – dependence and victimization; the centres to which many writers respond, with which they communicate, are no longer located in 'the West' but in Africa. Africa, in these contemporary visions, has always been the centre – as memory and cultural repository, but also as future world. This recentring of Africa is pertinent to the question of African literature – in its many permutations – and its relationship to 'Afrofuturism', the 'speculative fiction that treats African-American themes and addresses African-American concerns in the context of twentieth-century technoculture' (Dery 'Black to the Future': 180). Yet, as we continue to forge Africa's futures – literary and otherwise – and as we are reminded of the imperatives of decolonization, both within and outside of the academy, 'Afrofuturism' must also be interrogated for its relevance and pertinence to conversations about Black futures outside of the United States. In the move to release ourselves from one ill-fitting paradigm for African writing, we might be too hasty in adopting another one.

'Afrofuturism', the label given to the creative output of people racialized as 'Black' in the United States, was first used by white cultural critic Mark Dery in the introduction to his collection of interviews,

'Black to the Future' (1994). Dery limits his enquiries about science fiction to African-American writers, whom he identifies as members of 'a community whose past has been deliberately rubbed out, and whose energies have subsequently been consumed by the search for legible traces of its history' (180). In his search for the 'largely unexplored psychogeography of Afrofuturism' (187) Dery limits himself to interviews with writers born and raised in the continental United States; his only fleeting allusion to a Blackness that does not originate there is his brief mention of Lee 'Scratch' Perry, a Jamaican musical producer who is considered a pioneer of Reggae and Dub. Like Samuel Delany, Greg Tate and Tricia Rose, Scratch shares a history of enslavement; yet, unlike them, he is not considered for Dery's psychogeography. As if to reinforce Dery's myopia, Perry's music is described, 'at its eeriest … as if it were made of dark matter and recorded in the crushing gravity field of a black hole' (182). For Dery, Afrofuturism outside of the United States was impenetrable, incomprehensible.

Even in the interviews with the writers whom Dery classes as 'Afrofuturist', we see tension between his understanding of such an expansive concept as 'the Black experience' and those of his interviewees. In 'Racism and Science Fiction', Delany challenges Dery's reading of the Rastafari in William Gibson's *Neuromancer* (1984), as well as a continuing academic interest in cyberpunk as 'a largely nostalgic pursuit of a more innocent worldview' that is no longer valid (if it ever was) after the Civil Rights struggle in the United States (200). Tate reminds Dery that, 'for a lot of black writers, that desire to know the unknowable directs itself toward self-knowledge' (210), and that one can be 'backward-looking and forward-thinking at the same time' (211). Rose ('Lagos: Hope and Warning') warns against 'a construction of black culture in such a way that blacks [*sic*] are romanticised', as 'these visions constitute an extraordinarily contradictory reading of black creativity' (215). These critiques of Dery's definition notwithstanding, however, the 'Blackness' that was imagined in these interviews is overwhelmingly, if not completely, one only possible in the United States. With the United States' economic, military and cultural ascendancy in Africa and other global locations, the imperative remains to relocate our frames of reference. We cannot continue to expect the master's tools to dismantle his house – or indeed build our own.

Critical and creative ambivalence to the term 'Afrofuturism' has been noted and well documented – particularly in the statements of Nnedi Okorafor. In a 2014 interview at the University of Texas

(Austin)'s Symposium for African Writers, Okorafor and fellow author Sofia Samatar critiqued the idea that 'speculative' fiction cannot be 'realistic', and thus challenged 'rational' ('Western') notions of what is possible in genre fiction – indeed, they challenged the very formation of these generic distinctions. Okorafor noted that in many African cultures,

> the idea of the mystical being part of the world – the mystical and the mundane being combined – is a natural thing. ... When I first started writing, I was writing *memoir*, and I was digging into my own past and writing about things that had happened to me, and the fantastical aspects of those stories naturally occurred there ... For some of us, it's just the way we see the world. Also, the idea of separating science fiction from realism, it's like separating the present from the past and the future. Those aren't separated, those are all combined, so I don't see how you can look at science fiction and say, oh, that's not realistic fiction. (Jalada 'Things to Come', original emphasis)

As I am not the first to argue, the 'fantastic' and the 'real' are not at odds in African creative and cultural expression. Africa is no stranger to witches, wizards and warlocks who fly by night; it is not inherently 'bizarre' that the spirits may walk among us, or that they may be fallible. Amos Tutuola's *The Palm-Wine Drinkard* (1952) and *My Life in the Bush of Ghosts* (1954) follow their protagonist into the world of the dead and into forests inhabited by spirits, respectively – the latter begins when the narrator is separated from his brother in an attempt to escape slave raiders. The concept of 'alien' invasion is not altogether new to Africa either: Chinua Achebe's *Things Fall Apart* (1958) very realistically describes invading colonizers as 'locusts', and as 'white men who made the powerful guns and the strong drinks and took slaves away across the seas' (102–103). Olaudah Equiano described in 1789 his 'astonishment' and then 'terror' upon seeing a slave ship for the first time, and his firm belief that he 'had gotten into a world of bad spirits, and that they were going to kill me' – a belief inspired by their skin, hair and unrecognizable language (24). These examples demonstrate that key concepts of science fiction – first contact, invasion, extra-terrestrials, conflict, environmental destruction, alternative worlds, transmogrification, shape shifting, time travel – all have African precursors. Moreover, as colonized subjects, Africans well understand alienation, whether it be through transportation to a new land or, as in Okonkwo's case in *Things Fall Apart*, having your own land change irrevocably around you. African writers may well excel at science fiction, as it is the only genre expansive, flexible and 'unrealistic' enough to enable them (and us) to imagine otherwise. As Kodwo Eshun noted in 'Further Considerations of Afrofuturism' 'imperial racism' (287) has too long overdetermined Black Atlantic cultural production as backward-glancing – has exiled it from 'the field of the future' (288).[2] Reclaiming science fiction from Europe and the United States allows for release – not necessarily synthesis – out of the dialectical posture of writing back to the West.

In 2019, Okorafor coined the term 'Africanfuturism' to describe her own writing, and to critique the uncritical proliferation of Afrofuturism as a generic label. While she recognized that Africanfuturism is a 'sub-category of science fiction', she argued that it differs from Afrofuturism in that 'Africanfuturism is specifically and more directly rooted in African culture, history, mythology and point-of-view as it then branches into the Black Diaspora, and it does not privilege or centre the West'. Africanfuturism, therefore, is a contemporary, continental African phenomenon (even as it may exist, as Nnedi herself does, in Africa's diaspora). Africanfuturism is inward looking; it does not concern itself with writing back but instead, writes itself *in*. Africanfuturism, she continues,

> is concerned with visions of the future, is interested in technology, leaves the earth, skews optimistic, is centred on and predominantly written by people of African descent (black people) and it is rooted first and foremost in Africa. It's [*sic*] less concerned with 'what could have been' and more concerned with 'what is and can/will be.' It acknowledges, grapples with and carries 'what has been'. (Okorafor 'Africanfuturism Defined')

Of note here is that the Blackness (as self-identification) that Africanfuturism envisions does not exclusively position itself from the traumatizing event/s of enslavement. Africanfuturism, so defined, has different relationships with 'the past', and with ancestry, than do narratives of and from what is now called the 'New World', the Americas. History in Africanfuturist narratives does not have to be 'uncovered' or 'made legible' in the same ways; the futures that are imagined are less 'impossible', as they are known to have existed. Africanfutures are rooted in tangible pasts that pre-exist Anglo-European colonization, and as such do not seek to reverse the 'social death' of slavery (Patterson, *Slavery and Social Death*). Selfhood does not have to be reinvented in this imaginary as, unlike Afrofuturism, the Black people here are not 'strangers in a strange land' (Dery: 210).

In that same blog post, Nnedi also outlined her definition of 'Africanjujuism', 'a subcategory of fantasy that respectfully acknowledges the seamless blend of *true existing* African spiritualities and cosmologies within the imaginative' (Okorafor, emphasis added). Where Africanfuturism concerns itself with science fiction, Africanjujuism concerns itself with the supernatural. This supernatural, however, is not impossible; it is simultaneously forward-facing ('imaginative') and retro-rooted, employing as it does Africa's various 'traditional' religions and cosmologies – the ones that missionary Christianity has so long demonized. Again, unlike Afrofuturism, Africanjujufuturism does not have to invent new worlds – it can adapt 'old', existing ones. Africanjujufuturism tasks itself less with inventing a 'lost' past than it does with *reintegrating* its past(s) into its realities. The past is a living, breathing, adaptive tool of re/vision that can repeatedly be challenged, and evolve.

I want to turn now, to illustrate these reflections, to Nnedi Okorafor's 2014 *Lagoon* and Suyi Davies Okungbowa's 2019 *David Mogo, Godhunter*, two novels that, published five years apart, imagine twenty-first-century Lagos as the site of an alien invasion. *Lagoon* charts the first few days of an alien invasion, and we follow the protagonists as they bring the alien ambassador to their leader, Nigeria's president, in order to usher in a new Nigeria. In *Godhunter* the invasion is ten years old, and the aliens are Orisha (and lesser godlings) who have been cast out of Orun as the result of a divine civil war. David, who makes his living catching godlings, must become a god in order to stop the Orisha from completely destroying Lagos, and make it safe for humans to inhabit again.

Situating alien invasion, one of the oldest science fiction tropes, far away from New York, Tokyo or even London not only exemplifies the irrelevance of the 'West' to Africanfuturism's imaginative project, it also centres Lagos – Nigeria, Africa – as a site in which things happen, a place to which people (aliens, gods) want to come, rather than leave. In both of these novels the extra-terrestrials want to make Lagos their *home* – they are invested in it and want to settle there. This re-siting challenges established paradigms of emigration away from the 'global South' and isolation (indeed alienation) in the frozen north. The future that Okungbowa and Okorafor offer is located, real and recognizable; it is urban, African, pidgin and cosmopolitan. In addition to that – because of that – it is also retroimaginative; in these futures the past, in the form of spiritual/religious knowledge, comes to life and carries Lagos forward. I contend that these Africanfuturist novels, in bringing the spirits back to Lagos, repurpose religion under the guise of the alien invasion motif as an organizing, future-facing framework for imagining ourselves out of our very contemporary 'post'/colonial malaise.

Okorafor herself is joined by fellow writers NK Jemisin, Samuel Delany, and Nalo Hopkinson in decrying racism in science fiction, and critics such as Jessica Langer have recorded the same (Okorafor 'Lovecraft's Racism'; Flood 'NK Jemisin'; Delany 'Racism and Science Fiction').[3] As Langer states, in science fiction 'otherness is often conceptualised corporeally', and alienness 'often dovetails with the colonial [racist] discourse of the Other' (*Postcolonialism and Science Fiction*: 82). At the opening of the twentieth century, W.E.B. Du Bois described being Black as being a 'problem' (*The Souls of Black Folk*: 1), as being 'an outcast and a stranger in mine own house' (2). Authors like Okorafor and Okungbowa, writing in and of Africa in the twenty-first century, do not imagine a Blackness that has this alienating and dehumanizing relationship with whiteness; their characters, Nigerians at home in Nigeria, are hardly described as Black at all. In *Godhunter* and *Lagoon*, Africans are not monstrous – they do not need to be subdued by Horace Holly types.[4] Here and now, Africans are simply human beings who learn to accommodate un-human others in their own homes – homes that, because they are historically rooted and routed through precolonial African philosophy, can, will and do adapt to (incorporate) these futures yet to come.

The Lagos these texts imagine is not far off from the present; these futurescapes are familiar. Metropolitan Lagos, writes Rita Nnodim in 'City, Identity and Dystopia', 'astounds the senses' with its vibrant mix of cultures and languages (321). Okungbowa's David, half-god, half-human, experiences Lagos through his hyperkeen senses. Lagos' admixture is its own alchemy, and its inhabitants repeatedly renegotiate themselves and their identifications in ways that are constantly changing – survival in Lagos hinges on adaptation. Okorafor's aliens are, first and foremost, agents of change; her protagonists – Adaora, a marine biologist, Agu, a soldier, and Anthony, a musician – all have special abilities that resurface as a result of and for the purposes of the invasion. Lagos is also a space of hybridity: David, who feels at home nowhere, makes his home in what he calls the 'ugly' part of Lagos, Ìsàlè Èkó.[5] The bottom of the island – as Ìsàlè Èkó is literally translated – is the epicentre of not only this Falling but that of the Europeans in the fifteenth century (12).[6] Ìsàlè Èkó is a liminal space, which divides the mainland from Lagos Island proper. Perhaps it is not by accident, either, that Okorafor's aliens first approach Lagos from its waters. Lagos is also a site of migration, 'a large lab', as one recent immigrant described it, 'where you can experiment, pretty much freely', even as the quality of life in that city might be precarious (Rosen 'Lagos: Hope and Warning'). 'Third wave' fictions of Lagos, Nnodim continues, 'open up multiple relationships between urban landscapes and identities' (322) – Lagosians at once define and are defined by their space/s, even as those spaces may be contingent and unpredictable. Lagos is a city both 'postcolonial' and postmodern: if futurity is to be found anywhere, it can be found in Africa's largest, most populous, perhaps most chaotic city. Okorafor's aliens choose Lagos because in it they see 'love, hate, greed, ambition and obsession ... compassion, hope, sadness, insecurity, art, intelligence, ingenuity, corruption and violence. This is life', their ambassador says, and 'we love life' (112). It is in Lagos, too, that two of the gods, at least, select to be the site of their new Orun. In both narratives the city absorbs all that is thrown in and at it; Lagos' cacophony makes it a crucible for reinvention, 'the site of a multiple, sprawling heterotopia' (Nnodim: 322). Its inhabitants, ever adaptable, invest the city with meaning – their *faith* keeps it alive and regenerating.

What is most striking, in comparing these two novels, is that Lagos' sights, sounds and landmarks are recognizably real. Early in *Godhunter*, David uses Google Maps to locate his quarry; for those of us unfamiliar with Lagos, Google Maps makes an indispensable reading companion to Okungbowa's narrative. In *Lagoon*, Bar Beach, formerly a site of public executions and multiple floodings, is the epicentre of an imagined future city; it is here that Ayodele, the alien ambassador, first takes tangible shape. The Lagos-Benin Expressway comes to life as the Bone Collector, a spirit that is finally appeased when one of the aliens sacrifices herself to it. What is also quite real about these future Lagoses is that the city 'is not only inscribed with the shattered dreams of its inhabitants in spaces of destruction but also figures as a site where the postcolony stages its regime of domination and violence, which is frequently inscribed by features of the obscene and the grotesque' (Nnodim: 323). It is not by accident that Adaora and David must overcome the frustration of relying on state apparati – in the form of the police, military and civil servants – that are more unwilling than they are unable to effect necessary change. David, upon returning from his first job in the narrative, is extorted for 40 per cent of his fee at a police checkpoint (25–28). In *Lagoon*, military chauvinism results in Adaora's eight-year-old daughter being shot in her arm. In

Godhunter, Nigeria's central government is nowhere to be found. These future imaginings of Lagos are very aware of their present; they do not escape from them into an implausible utopia, or despair of them in an indulgent dystopia. 'Future', here, does not have to be implausible.

ALIENS DON'T HAVE TO BE EVIL: A CRISIS OF LEADERSHIP IN *LAGOON*

In *Lagoon*, Lagos descends into chaos and mayhem with the arrival of shape-shifting aliens who catalyse humans' ugliest impulses: Agu comes to Bar Beach after countermanding his superior officer in an attempt to prevent him from committing rape (Agu is badly beaten for his trouble); Adaora comes there after her husband, under the influence of a predatory 'pastor', hits her for the first time. Lagos' topography – its very name – make it ideal for this extra-terrestrial invasion, as the unnamed space aliens choose the water as their base. They make first contact with the marine creatures (and clean the water, at least for the non-human life forms) before they come to land and; with their technology, they put an end to Nigeria's dependence on crude oil. Yet, even in this advanced future, Ayodele (as Adaora names her) is introduced with terms from the past. Ayodele is first described as 'smoke', but the first person to see her – a mute child who later dies in the ensuing chaos – describes her as a 'water spirit' (13). The second person to see her, a sex worker who later becomes 'one of the loudest prophets of doom in Lagos' (15), describes her as a 'shape-shifter … devil' (14). It is not difficult to recognize, in these descriptions, the allusion to Mami Wata, avatars of whom exist throughout the West African diaspora. Mami Wata is a water goddess responsible for economy, trade and fertility. She, like her water element, is a conduit – slippery and elusive. Mami Wata is known for her 'irresistible seductive presence' (Drewal et al. 'Mami Wata': 61), which is made manifest towards the end of Okorafor's narrative when she resurfaces to take Father Oke, the predatory preacher, away with her into the Lagoon. Mami Wata, furthermore, has always been exoticized and othered. Her devotees, Henry Drewal asserts, reinvent and resymbolize themselves and their place in the world culture through the performance of foreignness ('Performing the Other': 160).[7] It is through Mami Wata's dissimilarity to themselves that West Africans came to understand and incorporate otherness. These Lagosians, Nigerians, Africans already have a frame of reference through which to negotiate the 'alien'; a frame that is by nature adaptable and future-oriented, future-engaged. Mami Wata, like Ayodele's species, is an agent of change, trade and exchange – Africa has *been* ready to absorb the future.

What this and other elements in *Lagoon* as well as *Godhunter* demonstrate is a manifestation of what Harry Garuba calls the 'animist unconscious', a collective subjectivity that 'operates through [a] continual re-enchantment of the world'. This continual re-enchantment (which, again, evolves and adapts) does not have to be religious but, in it, 'magic' – or 'magical thinking' – is

> not displaced but, on the contrary, continually assimilate[s] new developments in science, technology, and the organisation of the world within a basically 'magical' worldview. Rather than 'disenchantment,' a persistent re-enchantment thus occurs, and the rational and scientific are appropriated and transformed into the mystical and magical. ('Explorations in Animist Materialism': 267)

The mystical, the sacred, the supernatural – in the forms not only of Mami Wata's multiple avatars but of Ijele, the chief of masquerades, the Bone Collector, Udide Okwanka, and even Legba, not to mention the various Orisha that converge on Lagos in *Godhunter* – are made real in these narratives. Moreover, they are mapped onto Lagos' future as the logic that catalyses and shapes this modern metropolis to come. Once the gods land, there is no return to 'normal'.

The second part of *Lagoon* – entitled 'Awakening' – opens with a tarantula, relative of 'the great spider artist of the land, she who weaves all things into existence' (14), being crushed to death on the Lagos-Benin Expressway. This tarantula has felt the vibration of the alien landing, and knows that it is 'a call for change' (119). He does not get to see this change, unfortunately, but the prologue discloses knowledge that even Ayodele does not have: 'that there are other things inhabiting Lagos besides carbon-based creatures. There are greater beings of the earth, soil, sea, lagoon and land' (120). Here, Okorafor intimates that Lagos, Nigeria, Africa has knowledge, history, and culture that is older and deeper than these aliens, that precedes and will outlast – once again, absorb – them, as it has absorbed countless others. This absorption is never passive but alchemical, regenerative. This synchrony of the past working to sustain and direct the future to come – this is Africanfuturism. The great spider is Udide Okwanka (cousin of the trickster-storyteller Anansi, of the Ashanti and their descendants in the Caribbean), about whom Okorafor has written before. In her story 'Spider the Artist' (2011), Udide 'lives underground where she takes fragments of things and changes them into something else'. Udide, master of bricolage and reinvention, is right at home in Lagos, a city that collects and recreates people, cultures and languages from all over the world. Udide returns to 'welcome' us, near the end of *Lagoon*, to this recreated Lagos, thus locating Okorafor's narrative in a much larger, older one. Udide announces herself as 'the narrator, the storyweaver' who weaves all of these characters' stories, and indeed the story we are reading (228). She has been here before the aliens and will outlast them, but even she cannot see all things – the storyteller is not omniscient. She too has been enriched by the aliens, and is now 'stronger than ever' (229). Indeed, Udide is modern: in the midst of the chaos she reminds us that 'the modern human world is connected like a spider's web' – like her stories (194). The internet, a feat of modern technology, cannot pass her, the keeper of 'traditional' knowledge. The mystical trickster-artist is cemented in this narrative as not simply part of the fabric of this fictional tale, but as the *creator* of this and all stories. One cannot dismiss Udide and relegate her to the past because the future, *sef*, is in her claws.

Udide's connectivity and trickery are mirrored by the speaker of the 'Code Name: Legba' chapter, in which a 419 artist testifies the end of their career in scamming after Ijele, Chief of Masquerades, descends onto an internet café. Legba, of course, is an avatar of Esu Elegbara, an Orisha of roads, paths, and crossways – of wires and communication. Legba is also a trickster; he is believed to speak all human languages and facilitates (sometimes frustrates) communication. Okafor's earthly Legba (we do not have their real name, or gender) is quick to emphasize that what they see is not a man dressed as Ijele, 'but the real thing … over thirty feet high … moving. Alive'. 'The creature', as they describe Ijele, not only moves, but makes music (199). For Igbo people, Ijele 'references the importance of human and ancestral leadership [and] fecund spirit power … It embraces valued imported materials in its use of cloth, yarn, and mirrors' – again, tradition has always been dynamic in African cultures – 'and it exemplifies the dynamism of Igbo art in its stately, pulsating, radiant, and aggrandising performance' (Herbert Cole 'Igbo Art in Social Context'). He continues, 'each *Ijele* is a microcosm, a world renewed, and a tribute to ancestors, the gods, Igbo propensities for progressive change and the foreign, for fertility and renewal' (original emphasis). To Okorafor's 'Legba', an Igbo trickster masquerading as a Yorùbá Nollywood film director masquerading as a Pan-African sacred profaner, the music Ijele plays is 'the sound of life, the beginning' (199). In the shadow of Ijele, Legba vows never to scam again. The masquerade and one of the aliens – in the form of an elderly man – dissipate and enter a computer, and the criminal feels 'this great swell of pride and love' for Nigeria. 'I would die for it', they claim, 'I would live for it. I would *create* for it' (200, original emphasis). Ijele, materialized on earth, reconnects this young person to their creative purpose and repurposes their communication. When Legba returns to their computer they find that all of their emails and contacts have disappeared: their slate has been wiped clean.

The connection between creativity and trickery is not to be taken lightly. Ivan Van Sertima referred to the trickster, in African diasporic traditions, as a 'transcendent criminal' ('Trickster, the Revolutionary Hero': 107) but also a 'revolutionary hero'. The trickster, he argued, is a figure perpetually in revolt, a taboo, an outrage. Nevertheless, such a figure is also 'free spirit, original energy, pure primal power ... Such a form can imaginatively go back to beginning, to the unstructured chaos of origins, in order to free the human psyche to explore new orders and forms' (104). In *Lagoon*, the criminal was always an artist, and the technology that we associate with 'modernity' has ancient models. The past returns to reset the future, and this modern story has deep roots, right into the belly of the earth, where Udide resides – where, for many Igbo, ancestors reside. Even Udide acknowledges that the cosmos is 'soily'. The story, in its entirety, is to be found in 'the always-mingling past, present and future', she says, 'in the powerful spirits and ancestors who dwelled in Lagos' (194). The future, in *Lagoon*, does not depart from, but reconnects with the past – it radiates *inwards* as well as outwards. Lagos, as it stands, as it has stood, already has everything the future may need.

When Ayodele is 'taken to the leader', Nigeria's president, she explains that her people are not evil (as tricksters are often thought to be), but they are change (as tricksters often are). They are technology, she explains, which Okorafor has long maintained 'is just another form of juju' (quoted in Whitted '"To Be African is to Merge Technology and Magic"': 209). Africanfuturejujuism takes as granted that 'the mystical and the mundane have always coexisted' (209) so, in this future Lagos, Ayodele is not an aberration. We see this when she tells the president that 'we do not want to rule, colonise, conquer or take. We just want a home' (220). Ayodele's assimilation into Lagos goes further when she sacrifices herself to mob violence and her body dissipates into a white mist that all of Lagos inhales (269). Lagosians are left imperceptibly changed, except for a craving for eggplants (Ayodele's favourite food), and the president, made healthy, announces that 'corruption is dead in Nigeria' (279). In this revisioning of alien adventure/science fiction invasion fantasy, the 'invaders' do not take from but *give* to the receiving society. Okorafor, in dialogue with *Things Fall Apart*, imagines a world of exchange without colonization, commerce without genocide. Both before and after the Middle Passage, figuratively speaking, *Lagoon* imagines humanity expanded, augmented and reset.

NOT SUCH A BAD JOB: FINDING FAMILY AND FREEDOM IN *GODHUNTER*

Corruption may be dead in *Lagoon*, but the Lagos of *David Mogo, Godhunter* has not quite arrived at this future yet. Its aliens – the Orisha of Orun – have been resident in the city for some ten years, but corruption has only intensified. Ìsàlè Èkó has largely been abandoned by the country's elites, who have barricaded themselves in their homes far from the epicentre of The Falling. The gods have returned but Lagos is a shadow of its former self: the bustling metropolis has become 'a dumping ground' for everyone who could not escape (12). Not much has changed in this future Nigeria, and David is highly critical of it; as Rita Nnodim observes, Lagos is depicted as a site in which 'notions of citizenship collapse under the general feel of disillusionment in the face of tyranny and endless repetition' ('City, Identity and Dystopia': 330). David and his elderly guardian, Papa Udi, remain on Lagos Island for sentimental reasons: Udi found infant David in a guava tree in the yard of Cardoso House, where they have both lived since, in a colonial-era house that, even before The Falling, had seen better days. David gets mixed up in the business of the Orisha because his roof has caved in, and neither he nor Papa Udi can find 'legitimate' work to repair it. In this Lagos, godhunters like David and wizards like Papa Udi are frowned upon for working with 'the occult' but have become necessary evils because the state – in the form of the woefully underfunded Lagos State Paranormal Commission (LASPAC) – has failed to protect its citizens from the menace of the godlings.

Like Okorafor, Okungbowa does not have to build a new world in order to imagine the future; the technology in *Godhunter*, though, is supernatural – juju. Like technology, the gods in *Godhunter* can be manipulated – their powers have been significantly weakened by The Falling, and their desires are very human indeed. We learn that Aganju, Orisha of volcanoes and the wilderness, along with his brother Sango, Orisha of thunder, desire to recreate a new Orun on Lagos Island.

Aganju was banished from Orun by Obatala for his taboo experiments in recreating (what he saw as expanding) the divine, manifested in the 'godlings' that now infest Lagos. Through his human proxy, Ajala, Aganju is kidnapping street children to continue similar experiments using the god-essence of Ibeji, the twin Orisha of 'Divine Abundance'. As Kehinde tells David, Ibeji are 'the keepers of the essence of all things. We are wisdom, growth, prosperity, fertility. We are life. We are existence' (50). Ibeji are casualties of the civil war that destroyed Orun as a result of Aganju's banishment; Kehinde continues, 'we did not ask for this war, but we became victims anyway. We are refugees in a place we cannot call home; we cannot go back to the place we *should* call home' (58, original emphasis). As much as *Godhunter* is a story about an alien invasion, it is also a modern – mysticized – story about migration, exile, displacement, and diaspora. Lagos is a city of immigrants, and diaspora is fast becoming a global condition –a 'condition of subjectivity' that, as Lily Cho argues, 'brings together communities which are not quite nation, not quite race, not quite religion, not quite homesickness, yet they still have something to do with nation, race, religion, longings for homes which may not exist' (14, 13). This is the futuristic, currently dysfunctional condition from which Lagos must negotiate being and belonging in the twenty-first century.

Aganju sees himself as a philosopher, scientist, scholar, and rebel, in defiance of Obatala's dictatorship: Lagos is the site of his utopic reinvention. As he says to his brother, 'we do not need to find our centres because we do not need to go back at all' (234). He believes that 'a few cannot dictate the undertakings of the many' (105) but irony appears lost on this high god as, in order to achieve his freedom – to make his home – he is willing to destroy all of Lagos, the home of many millions. Aganju and Sango represent another, even more dangerous military dictatorship, a form of governance with which Nigeria is well familiar. David must defeat them, but before he kills Sango he is left with that god's memories and impressions – 'the raw, unfiltered emotion [of] agonising pain, but underneath it sadness, the sadness of never getting to see the end, of never knowing what one has built' (234). Where there is war there is also sadness and, in Aganju/Ajala, David senses how the thirst for knowledge and the search for home have been corrupted. David recognizes this latter struggle in particular, as he does not know either of his birth parents. He knows his mother is a high god, but not to which pantheon she belongs. It is to her that he owes his special powers, but from her he has also inherited the feeling of being 'lost, abandoned by the people who're [*sic*] meant to love you most, left alone to fend for yourself in a strange place … to settle into a world that does not suit you' (43). When David's mother does materialize, we learn that she is Ogun, god of war. Ogun is normally imagined as male, but having her recreated here as not only female but a mother – albeit not a traditional one – complicates not only the generativity of motherhood but the destruction of war. Okungbowa's Ogun, like the other Orisha we encounter, has lost much of her power, as well as her taste for war; she is still well respected, but no longer feared. She, who has previously only been able to communicate with David in dreams, has come to reconcile with and pass the mantle to her son, who will use her ancient power to reclaim Lagos' present, as well as its future. As does Legba, Ogun, in *Godhunter*, belongs to several pantheons, to several believers; it is for this reason, chiefly, that she separated herself from her son. War is perhaps universal, and history must repeat itself in order to secure Lagos' future. In this narrative, though, this war will be on behalf of humanity.

If we consider the invasion of the Orisha to be the past returning, David can be considered a prodigal – diasporic? – son of Africa. He is able to find the mother that he lost, unlike those of African descent in the diaspora who must live with the rupture of the Middle Passage forever. Saidiya Hartman describes this diasporic rupture as 'losing one's mother': 'losing your identity, your language, your country … it's about those losses that haunt us, those ancestors who we know but can't name. We feel their presence but they're without names for us' (Chideya and Hartman '"Lose Your Mother" Author Finds Heritage in Africa'). This is David's relationships with his mother for most of his life, and most of this narrative. In fact, neither he nor his mother can remember – and so we never discover – his 'true-true' name, which she can only recall when she needs to utter it. Although keeping his name a secret – protecting the kernel of himself from those who may wish to do him harm – has kept David alive, it has estranged him from her; when mother and son are reunited, 'David' is the name she learns to use. David is not fluent in any of Nigeria's indigenous languages (and suffers for this in many instances); being half-god, half-human, he has never belonged anywhere. Severed from his kin, he too exists in the subjective condition of diaspora. Not until he is reunited with his mother can he really become himself: Amúnáwá, the Firebringer. It is through this trial by fire that gods and humans can begin to peacefully coexist in Lagos. In this sense, *Godhunter* hints towards incorporating Africa's diaspora into its future, but first, Lagos must heal itself, so that people can 'resume their business of being very human in one of the most inhuman cities in the world' (177).

CONCLUSION

Both of these novels demonstrate an understanding of the futuristic and fantastical as simultaneously natural, pre-existing. Beyond asking 'What if?' they ask 'Why not?' The futures they imagine are plausible, and take place in a Lagos that is familiar to our present. In so doing, they reconfigure – critique – the tenets not only of what we have come to understand as Afrofuturism, but postcolonial African literature more broadly. They remind us that 'Afrofuturist' and 'postcolonial' are not mutually exclusive, and that African literature exceeds and expands these generic labels anyway. In *Lagoon*, Lagos (Nigeria, Africa) is shown to have already had frameworks with which to read the future. In *Godhunter*, the supernatural (spiritual) is brought to earth, literally. Both narratives, though, render the mystical mundane, remind us, through the imaginative project of Africanfuturejujuism, that the pathways across space and time have always been open.

The inhabitants of these not too distant, not unfamiliar Lagoses must, in the wake of an other-worldly disaster, re-evaluate what it means to belong to one of the world's most chaotic, cacophonous, dys/functional, diasporic locations. Faced with the end of life as they know it, ordinary – and not so ordinary – Lagosians find a way to reclaim – and in so doing, recreate – their home using the technologies of their pasts as blueprints to their futures. Lagos, a place where existence is precarious and identity is in constant negotiation, provides the perfect alchemy for those who no longer have a home to forge a new one – through peaceful or violent means. 'The future', in these African narratives, will be fought for and won *at home* – there is no going back (or away), and what has been severed must be reintegrated. While set in the future, *Godhunter* and *Lagoon* are very mindful of Lagos', Nigeria's, Africa's 'post'-colonial present; in both narratives, corrupt governments are overthrown and humanity is re-membered. Moreover, Lagos is reimagined as *aspirational*, a city in which humans, sea creatures, gods, and aliens can and want to coexist. These are not utopias but neither are they impossible – the futures here imagined have always been real.

NOTES

1 The quotation in the chapter title is from Okungbowa, *David Mogo, Godhunter*: 177.
2 Eshun does not critique, as Okorafor does, Afrofuturism's US-centric geopolitics.
3 Nalo Hopkinson, interviewed by Jessica Langer, referred to racism in science fiction as being 'like the elephant in the room .... Actually, no; it's like there *should* be an elephant, but instead, there's an elephant-shaped hole', in Langer, *Postcolonialism and Science Fiction*: 1.
4 Horace Holly is the protagonist of H. Rider Haggard's *She* (1887), who travels into the 'lost kingdom' of the African interior.
5 David uses Lagos' original Yorùbá name, to emphasize the city's history before and beyond colonization.
6 'The Falling' is the term used to describe the invasion of Lagos by the Orisha.
7 Drewal builds on Roy Wagner's *The Invention of Culture* (1980).

WORKS CITED

Achebe, Chinua. 2001. *Things Fall Apart.* London: Penguin Classics.
Chideya, Farai and Saidiya Hartman. '"Lose Your Mother" Author Finds Heritage in Africa'. *NPR*, 23 January 2007. www.npr.org/templates/story/story.php?storyId=6955366 (accessed 8 January 2021).
Cho, Lily. 'The Turn to Diaspora', *TOPIA: Canadian Journal of Cultural Studies* Vol. 17, No. 11, 2007: 11–30.
Cole, Herbert. 'Igbo Art in Social Context', nd. https://africa.uima.uiowa.edu/topic-essays/show/15?start=13 (accessed 8 January 2021).
Delany, Samuel. 'Racism and Science Fiction'. In *Dark Matter: A Century of Speculative Fiction from the African Diaspora*, ed. Sheree R. Thomas. New York: Warner Books 2000: 383–397.
Dery, Mark. 'Black to the Future'. In *Flame Wars: The Discourse of Cyberculture*, Mark Dery (ed.). Durham, NC: Duke University Press, 1994: 179–222.
Drewal, Henry John. 'Performing the Other: Mami Wata Worship in Africa', *TDR* Vol. 32, No. 2, 1988: 160–185.
Drewal, Henry John, Marilyn Houlberg, Bogumil Jewsiewicki, John W. Nunley, and Jill Salmons. 'Mami Wata: Arts for Water Spirits in Africa and its Diasporas', *African Arts* Vol. 41, No. 2, 2008: 60–83.
Du Bois, W.E.B. *The Souls of Black Folk.* New York: Dover Publications, 1994 [1903].
Equiano, Olaudah. *The Interesting Narrative of the Life of Olaudah Equiano.* Moscow: Dodo Press, 2007.
Eshun, Kodwo. 'Further Considerations of Afrofuturism', *CR: The New Centennial Review* Vol. 3, No. 2, 2003: 287–302.
Flood, Alison. 'NK Jemisin: "It's Easier to Get a Book Set in Black Africa

Published if You're White"'. *The Guardian*, 4 May 2020. www.theguardian.com/books/2020/may/02/nk-jemisin-its-easier-to-get-a-book-set-in-black-africa-published-if-youre-white (accessed 13 January 2021).

Garuba, Harry. 'Explorations in Animist Materialism: Notes on Reading/Writing African Literature, Culture, and Society', *Public Culture* Vol. 15 No. 2, 2003: 261–285.

Haggard, H. Rider. *She*. London: Longmans, Green & Co., 1887.

Jalada. 'Things to Come' (transcript) by Aaron Bady, 15 January 2015. https://jaladaafrica.org/2015/01/15/things-to-come-transcript-by-aaron-bady (accessed 7 January 2021).

Langer, Jessica. *Postcolonialism and Science Fiction*. London: Palgrave Macmillan, 2011.

Nnodim, Rita. 'City, Identity and Dystopia: Writing Lagos in Contemporary Nigerian Novels', *Journal of Postcolonial Writing* Vol. 44 No. 4, 2008: 321–332.

Okorafor, Nnedi. 'Africanfuturism Defined', 19 October 2019. https://nnedi.blogspot.com/2019/10/africanfuturism-defined.html (accessed 7 January 2021).

—— *Lagoon*. London: Hodder & Stoughton, 2014.

—— 'Lovecraft's Racism & The World Fantasy Award Statuette, with comments from China Miéville', 14 December, 2011. http://nnedi.blogspot.com/2011/12/lovecrafts-racism-world-fantasy-award.html (accessed 13 January 2020).

—— 'Spider the Artist', *Lightspeed Magazine* No. 10, 2011. www.lightspeedmagazine.com/fiction/spider-the-artist (accessed 15 January 2021).

Okungbowa, Suyi Davies. *David Mogo, Godhunter.* Oxford: Abaddon Books, 2019.

Patterson, Orlando. *Slavery and Social Death: A Comparative Study.* Cambridge, MA: Harvard University Press, 2018.

Rosen, Armin. 'Lagos: Hope and Warning', 8 July 2018. *City Journal*. www.city-journal.org/html/lagos-nigeria-16011.html (accessed 14 January 2021).

Rushdie, Salman. 'The Empire Writes Back with a Vengeance'. *The Times*, 3 July 1982.

Van Sertima, Ivan. 'Trickster, the Revolutionary Hero'. In *Talk That Talk: An Anthology of African-American Storytelling*, Linda Goss and Marian Barnes (eds). New York: Simon & Schuster, 1989: 103–112.

Whitted, Qiana. '"To Be African is to Merge Technology and Magic": An Interview with Nnedi Okorafor'. In *Afrofuturism 2.0: The Rise of Astro-Blackness*, Reynaldo Anderson and Charles E. Jones (eds). Lanham, MD: Lexington Books 2016: 207–213.

Southern Africannearfutures

Black-tech, Ambivalence, & Speculation in Namwali Serpell's *The Old Drift* & Masande Ntshanga's *Triangulum*

JEFFREY G. DODD

AMBIVALENCE AND BLACK-TECH IN AFRICANNEARFUTURES

Storytellers from Africa and its diasporas have been presenting non-realist imaginative works as long as there were people to hear or read them. The social and (neo-/post-)colonial politics of how those stories are treated by historians and critics, and the labels used in discussing them, are an important element of how we understand the speculative vein(s) of African fiction(s). The contributions to this volume will analyse many texts from the nineteenth and twentieth centuries and pursue vigorous debates about the labels we might apply to them. For the purposes of this chapter, I will use the term 'Africanfuturism', as defined by its coiner Nnedi Okorafor, and I will pay specific attention to these elements of the sub-genre that Okorafor argues distinguish it from Afrofuturism:

> The difference is that Africanfuturism is specifically and more directly rooted in African culture, history, mythology and point-of-view as it then branches into the Black Diaspora, and it does not privilege or center the West.
>
> Africanfuturism is concerned with visions of the future, is interested in technology, leaves the earth, skews optimistic, is centered on and predominantly written by people of African descent (black people) and it is rooted first and foremost in Africa. It's less concerned with 'what could have been' and more concerned with 'what is and can/will be'. It acknowledges, grapples with and carries 'what has been'. ('Africanfuturism defined')

Okorafor importantly acknowledges that Africanfuturist texts grapple with distinctly African pasts and presents, while concerning themselves with potential futures.

Additionally, Okorafor's mention of technology invites a consideration of just *how* Africans' embrace of, and agency with, technology inform potential futures. Scholar M. Haynes considers this interest in technology to be a foundation for 'Black-tech', a sub-genre of works 'allowing for African Diasporic people to wrestle with science and the ways that it affects them'. Haynes elaborates:

> Texts that explore experimentation on African Diasporic people, using them as a form of human technology, or their contributions to science are right at home in this sub-genre. However, Black-tech also can have positive uses. Due to the distrust many African Diasporic people have of science, Black-tech can be used to show that with training and focus they can be in control of the very entity that oppresses them. Works that show African Diasporic people creating and utilizing technologies of today, or simply interacting with science positively are the types of empowering works that make up Black-tech. ('Black Speculative Fiction Definitions')

The ambivalence implicit in this definition, the knowledge that Africans' use of technology has and may yet be both a (neo-/post-) colonial instrument of oppression *and* an instrument to be used for their own potential liberation, is a primary subject of this examination.

Further to this discussion of definitions is my insistence on the phrasal terms (neo-/post)colonial and (neo-/post-)colony. These constructions strike me as most intellectually reflective of the overlapping and permeable social, political, historical, and geographical conditions experienced by the peoples and descendants of peoples impacted by European colonization. Attaching, without disambiguation, 'neo-' and 'post-' to 'colonial' reminds us that these experiences exist on a kind of spectrum that, from the third decade of the twenty-first century seems always and already compressed and most often indistinguishable. That is, it is impossible to separate the conditions of neo- or post-coloniality from the colonial stem, and the interplay between neo- and post- is more fluid than separate terms allow.

Ultimately, I argue that black-tech only amplifies the tensions of ambivalence and hybridity familiar to readers of African and other (neo-/post-)colonial literatures. Indeed, ambivalence and hybridity are central to the work of foundational theorists like Fanon, Said, Bhabha, Spivak, Young, and others, and form a key pillar of scholarship applied to literary texts from around Africa and its diasporas. Discussions of (neo-/post-)colonial ambivalence tend to emphasize the conflicted pasts and presents of (neo-/post-)colonial subjects. For example, Bhabha explains mimicry in the colony as a foundational expression of ambivalence: 'mimicry emerges as the representation of a difference that is itself a process of disavowal. Mimicry is, thus, the sign of a double articulation; a complex strategy of reform, regulation, and discipline, which 'appropriates' the Other as it visualizes power' (Bhabha 'Of Mimicry': 126). Along the way, imposition of regulations and discipline on colonial subjects reveals ambivalences, or in Bhabha's terms 'produce[s] its own slippage' (126). The colony and its 'othered' subjects will always comprise a fractured replica of the cultures and discourses manifest in the colonial encounter. While Bhabha suggests (in *Location of Culture*, for instance) that hybridity could represent a transcultural path forward, I take the less saccharine view shared by Young (*Colonial Desire*) and Spivak (*A Critique of Postcolonial Reason*). Even a hybridized (neo-/post-)colonial construction will be characterized by the tensions and enduring effects of racial, cultural, and social hierarchies rooted in colonial histories.

Thus, I consider black-tech an operative site of hybridity wrought with ambivalence because, while the development of technological interventions and developments by subjects of the (neo-/post-)colony may be a method of self-actualization and potential resistance, *it is also* a form of submission to technologies rooted in colonial power structures. If an Africanfuturist lens allows us to see into the future, in Okorafor's terms, to move beyond 'what could have been' towards 'what is and can/will be' ('Black Speculative Fiction Definitions'), the ambivalences revealed in the relationships between (neo-/post-)colonial subjects and the technologies they create provide starting points for identifying slippages in those futures. While such ambivalences and slippages are present in Africanfuturist texts exploring visions of distant futures, they are arguably more salient in those that imagine futures closer to our contemporary time. To highlight this distinction and focus in my analysis, I will use 'Africannearfutures' as a playful adaptation of Okorafor's term Africanfuturism. In do so I will take up, as a significant site of critical inquiry, the use and abandonment of technology in two very recent southern African speculative novels that imagine nearfuture versions of South Africa and Zambia: Masande Ntshanga's *Triangulum* and Namwali Serpell's *The Old Drift*.

COLLABORATION, AMBIVALENCE, AND AGENCY IN *TRIANGULUM*

Award-winning South African writer Masande Ntshanga's second novel, *Triangulum*, was published in 2019 by the Random House imprint Umuzi Books in South Africa and by Two Dollar Radio, an independent press in the United States. Reviewer Ian Mond, calling it a 'found manuscript' novel, draws comparisons to James Hogg, David Means, and Mark Z. Danielewski. The central premise of *Triangulum* is explained in its fictional Foreword. Dr Naomi Buthelezi is recruited in 2043 to assess the veracity of claims presented in a mysterious package delivered to the offices of the South African National Space Agency. The package comprises three artefacts, whose sender demands be considered an inseparable single document: a journal capturing the day-to-day life in 1999 of a teen living through her father's illness and death in Ciskei, a few years after her mother's mysterious disappearance; recordings from regression therapy sessions conducted in 2035, during which the same young woman recounts her and two friends' 2002 efforts to solve the riddle of three local kidnappings and find the source of her mother's vanishing; and an interlude, 'Five Weeks in the Plague' in which the same narrator in 2025 finds herself working on a shadowy government data-mining project before joining a group of hackers and ecoterrorists whohave grown disillusioned with the government's increasingly dangerous embrace of capitalism. Together, these testimonies correctly predicted a 2040 bombing on Table Mountain, and ultimately foretell Earth's potential destruction in 2050. Buthelezi's job is to determine whether there is enough truth in the forecast to warrant a governmental response.

In the book's middle movement, the unnamed narrator – a former science writer and child maths prodigy, who throughout her young adulthood saw visions of triangles, which she called The Machine – has been working as a data analyst for a private social welfare firm aligned with the South African government 'to monitor the rise and decline of the worker population in the metropolis and beyond, and to note their consumption patterns in the townships and the CBD' (209). The narrator and her colleagues work with the implicit assumption that their labour is contributing to an 'Urban Renewal Project' that will create residential zones. Just as natural resource mines constructed and exacerbated social inequities for South African miners in the twentieth century, the embrace of data-mining, a peculiarly twenty-first-century extractive technology, is being used to create similar inequities. If ambivalence is characteristic of the (neo-/post-)colonial subject, that ambivalence has not yet left our nearfuture narrator: 'I sometimes saw myself as a different woman, one from seven or eight decades earlier: a worker at the Social Welfare Department at the non-European Section of Johannesburg, driving to Alexandra and handing out surveys to families whose children were bedridden as a matter of course, their bellies swollen taut with kwashiorkor' (210). This narrator is clear-eyed about her role as collaborator, just as she is aware that her parents were complicit with the apartheid government decades before her. Indeed, multiple times in the novel, she is spurred to recall her parents' own conflicted positions as administrators for the apartheid-allied government of Ciskei.

However, her complicity is not unreflective or even uncomplicated. Three years earlier, a failing government healthcare system left her vulnerable to losing necessary medication, which in turn made her a target for a corporate hacker, M/A/R/K, who convinced her to exploit a backdoor in the agency's network. If she was not convinced by his contention that '[i]t's closing – our window of opportunity to do something about our totalitarians' (231), the effective pharmaceuticals he offered were compelling enough to convince her to install a worm that infiltrates the network. Thus, both our tech-savvy, data-analysing narrator and a black corporate hacker are willing and able to attack the neocolonial data-mining operation that was instrumental in the country's 'economic shift toward information trading' (223).

These ambivalences, M/A/R/K's ability to exploit weaknesses in corporate systems while working from a Department of Defense grey site, our narrator's willingness to both work for and disrupt the agency, and the constant reminders to her of her parents' work for the apartheid-aligned government of Ciskei, represent constant tensions for black South Africans. Ntshanga is well aware of the ongoing conflicts and ambivalences the legacies of apartheid create. One of the book's epigraphs anticipates this theme, as readers are pointed to Leonard Thompson's claim in *A History of South Africa* that '[m]ost of the black bureaucrats, numbering around 820,000, were reliable servants of the regime on which they depended for their livelihood' (quoted in *Triangulum*: np). The novel's narrator, then, is in many ways living the legacy her preceding generations established: she is stuck in serving an undignified system that offered unsuitable provision in exchange for unthinkable oppression.

In fact, Ntshanga has explained that such conflicted collaboration appears to have been a key part of the novel's composition:

> I realised, as I was writing, that South Africa's dystopian future could easily be imagined through its dystopian past. In the sense that we've already been through it – the kind of tiered and oppressive technological society that's a trope in a lot of science fiction narratives about the future. Including human experiments and large-scale environmental disasters. The present is where the two converge, since South African

> history occurs as a continuum (with apartheid overlapping into the post-apartheid moment and so on) instead of as separate, conclusive chapters. In other words, *in order to imagine what could be, all I had to do was research what had been and crib it.* (Ndlovu 'South Africa's Dystopian Future', interview with Ntshanga, emphasis added)

This recognition is, again, presaged by another of the novel's epigraphs, the maxim attributed to Kobo Abe: 'The Future is forever a projection of the present' (np). In this sense, *Triangulum*'s speculative treatment of postcolonial ambivalence is a mere extension of what scholars like Matthew Omelsky and Claire Ducournau have observed with respect to other African authors and texts: the conflicted consciousness is nearly intractable due to the allegiances and dependencies created by (neo-/post-)colonial project(s).

However, *Triangulum* does invite us to look forward, to recall Okorafor's 'visions of the future', and move beyond mere ambivalence in two important ways. A shift occurs as the narrator's work with the agency becomes more fraught. She is selected to be part of a special team on a new quasi-governmental project. It is 2025 and she will shift from analysing decision-making and consumption patterns of Johannesburg's worker population to a decidedly more sinister project. Her agency has implanted data chips into a series of test subjects who will be surveilled, their behaviour and emotional responses analysed to 'decode human desire' (243) so that, '[i]t will be within our power to influence them toward better choices' (243). Part of their work is to monitor videos of the test subjects' daily lives. When the team watches their third patient commit a murder, the narrator quits her job. In the succeeding weeks she shadows a friend who is an organizer for an ecoterrorism network, The Returners, and eventually uses her connections and tech savvy to support a mission to destroy a paper company. During the mission, she sees The Machine again, this time with a clear overlay of three natural areas in the Eastern Cape and including a vision of her long-disappeared mother (317–318). It is, then, possible to eschew static ambivalence, abandon complicity, and actively resist the structures contributing to the ambivalence in the first place. In this move, Ntshanga offers a vision of newly uncomplicated and markedly unambivalent resistance.

Ten years later, in 2035, she discovers that The Returners were led by her mother, who believed that she had been communicating with an alien signal. The Returners' research suggested that humans are responsible for Earth's degradation, which has had 'a ripple effect on how the universe balances itself' (328). The narrator's mother had all along felt that she had needed to prioritize the work of saving everyone, even at the expense of losing her daughter (329). The narrator eventually realizes that she must visit the Vredefort Crater site and recognizes the reality that she will be responsible for sharing the civilization-saving message of imminent destruction (337). This realization, the reader is led to infer, causes the narrator to collect the manuscript – her childhood notebook recalling her parents' navigating apartheid and her mother's disappearance, the recordings of her regression therapy, and the recounting of her work for the government agency – and to share her conclusion with the South African National Space Agency: if significant changes are not undertaken to correct for Earth's imbalance, an interstellar object will destroy the planet in 2050. This action of alerting the Space Agency represents a small but significant moment of agency for the narrator. She is letting them know that her embrace of technology and former complicity will not continue; that she knows the violence created by the pseudo-governmental Urban Renewal efforts, which later became labour camps; and that she can defiantly and vocally ask for better.

Finally, the novel's 'found manuscript' frame comes back into view. In the Foreword, Dr Naomi Buthelezi tells her readers:

> the instruments of power could not be trusted with our future. Instead, our aim is to deliver it to the population, even though, we do not know how events will unfold, and whether or not we will survive. The choice is ours. (11)

In this future, 2043, ambivalence is gone. (neo-/post-)colonial governments have been exposed as frauds, and those who trace their lineage to the time before European conquest must take control over the potential futures available to that lineage. Now disabused of the blind promises of technology, indigenous South Africans might pick apart those old ambivalences and declare a new re-balancing in hopes of saving themselves and the world.

AMBIVALENT BLACK-TECH AND RE-BALANCING IN *THE OLD DRIFT*

Early in the extended third section of Namwali Serpell's celebrated 2019 novel, *The Old Drift*, Joseph and Jacob have an uneasy truce in the carpenter's courtyard in Kalingalinga, a hardscrabble neighbourhood adjacent to the University of Zambia, out of which Joseph has recently dropped, and into which poverty-stricken Jacob could never hope to gain admission. Joseph is trying to help fulfil his late father's plan to use CRISPR gene editing to cure 'The Virus', while Jacob is cracking his way through the prototype and design of microdrones. One day in 2016, Joseph notices Jacob's new Digit-All Bead, a small implant allowing the hand to function as a digital device:

> Jacob crooked the fingers of his left hand at the top knuckle, casting a square of light onto the palm. He tapped this 'screen' with his right index finger. The Google home page appeared. 'See?' He extended his hand palm up. 'Even us poor people, we have Googo now'. (424)

In this courtyard, with one young Zambian on the cutting edge of gene editing and epidemiology, and another cutting his teeth on biomimicry and micromechanics, we see a tech implant illuminate a hand for a web search.

Perhaps more obviously than Ntshanga does in *Triangulum*, though not necessarily more importantly, Serpell portrays black-tech as a central means by which Zambians might navigate their (neo-/post-) colony. In the scene recounted above, we recall Haynes' description of black-tech and we see the ways in which Zambians' facility with, and creation of new, technologies represent opportunities for oppression and liberation. Jacob angrily believes that Joseph's father's work may have infected his mother with The Virus (429). At the same time, Jacob feels freed when he is able to make a connection between his grandmother's starring role in the abortive Zambian Space Programme and his own skills: 'Who knew technology was a family tradition – in his very blood! For the first time, Jacob could see a connection between his hands and his mind, and it was precious' (444). In time, Jacob creates a lightweight, mosquito-inspired drone wing, and constructs a system where multiple drones could move together and (if necessary) sacrifice members for fuel. 'It would be a swarm that ate itself once in a while to stay afloat' (483). This operation, full of Jacob's typical ingenuity and representative of local problem-solvers in various parts of the continent, secures him a windfall buyout from a Sino-American panel of investors and corrupt general who sees these 'Moskeetoze' as a powerful tool to enhance his power over the people (484).

Thus, ambivalence again. The Virus research, if it is to go forward, must be tested on humans, and those humans probably are not going to volunteer. Multiple times, the characters remind one another that African bodies have been used without consent for experimentation. Jacob's microdrones secure his future, but no one yet knows the purpose to which the General will put them. Even the Bead, the digital device implanted by way of the median nerve that is well suited to developing countries, where power cuts, poor schools, and weak communications infrastructure combine to make ready markets, is poised to become a tool of state surveillance. This ambivalence is amplified when Jacob shares his second generation of microdrones with Joseph and his girlfriend, Naila. Naila is initially put off, to which Joseph replies that progress is inevitable. '"Progress?" she said. "Progress is just the word we use to disguise power doing its thing"' (506). Naila to this point has certainly appreciated technology when it works in her favour, as the Bead did when she visited India to return her father's ashes. But she is also wary of its power and sceptical that Digit-All free Bead giveaway was anything more than a test of how well the product worked with blacks' melanin, repeating the familiar line '[i]f the product is free, you're the product' (507) and questioning the moral structures inherent in any technological instrument (508). All this while her job at the Department of National Registration, Passport, and Citizenship's 'Electronical Administration' office includes responsibility for Beading persons seeking registration card renewals (510–511).

Naila's clear ambivalence about her technology, about her own complicity in the (neo-/post-)colonial administration mirrors that of *Triangulum*'s narrator. In this case, however, the narrative is chronologically linear, and Naila's response more obviously discernable. She becomes increasingly sceptical of the Bead's powerful surveillance capacity. As she demonstrates increasing militancy on the issue, she is convinced the government will try to use the Beads to force citizens to take a test vaccine (524). When a friend hacks the Digit-All system to confirm that the Beads are likely being used to monitor vaccine compliance (530), the next move is perhaps predictable when we recall that this trio was descended from women revolutionaries: Naila's grandmother had advocated for the Tonga who were displaced by the construction of the Kariba Dam, Joseph's was a university Marxist in the years just after independence, and Jacob's 'gogo' was a direct combatant against the British during the colonial resistance effort and had recently been thumbed for burning a government vaccine clinic. With that set of legacies, the trio launches a grassroots anti-government, anti-vaccine movement, SOTP (Sum of the Parts), for which the first major rally is attacked by a police force using microdrones that inflict dozens of small bites that Naila is convinced have non-consensually delivered the vaccine to attendees (543–544). In this moment we see what might have been the peak of ambivalence regarding both resistance to the (neo-/post-)colony and the embrace of black-tech. With respect to resistance, the SOTP gathering, neither protest nor riot, was co-opted to achieve the government's goal of forcing vaccine compliance. On the technology front, a central switch was flipped to turn off all attendees' Beads so they could nt be used to illuminate the drones attacking them, the technology Jacob created. We see an echo of *Triangulum*'s attention to the narrator's collaboration with her government. In both cases, the technology created or mastered by the (neo-/post-)colonial subject is used to advance the state's oppressive efforts.

The microdrone attack on the people allows Jacob to recognize that the Beads are central to the government's ability to enforce its policy, and that they are controlled centrally. By this point, the tail-eating snake is clear: the embrace of technology means that technology must be resisted by technological means. The novel's climactic moment occurs when Naila, Jacob, and Joseph identify the central Bead server's power source at Kariba Dam, itself a one-time technological marvel, the construction of which destroyed the Tonga people's homeland and the broader ecosystem of the central Zambezi basin. After years of poor maintenance and misused international aid (Serpell 'Learning from the Kariba Dam'), the dam is widely known to be in danger of a potential catastrophic failure, and this fragility is key to the novel's final unfolding. The trio's attempt to disrupt the Bead server unintentionally causes the dam's failure. By the novel's end, the Kariba Dam has failed, the Zambezi basin is inundated, 'the whole country was drowned' (563). Again, technology's possibility has brought with it unimaginable consequences.

ESCHEWING TECHNOLOGY IN SOUTHERN AFRICANNEARFUTURES

As speculative texts, these two novels could easily be seen as harbingers of doom, and some readers will surely see them as dystopic warnings. Though *Triangulum* delivers a message of destruction, two elements of that message are noteworthy.

First, it offers both a diagnosis and a prescription: 'If the corporatization of Earth does not end, then humankind will advance, but it will destroy this planet. It will seek other worlds and its imbalance will infect and spread, disturbing the universe and its calibration … before the universe rectifies itself and regains its balance' (335). Just such a re-balancing had been sought by the amaXhosa for generations, most notably represented by the cattle-killing episode of the nineteenth century (recounted in *Triangulum* and featured in Zakes Mda's *The Heart of Redness*), and the need for it is here is centred for consideration by all peoples of the Earth.

Second, its message is controlled by Africans, and the power to respond is in the hands of its African audience. As noted above, the found manuscript frame invests control of delivering the diagnosis and prescription in Naomi Buthelezi, whose lineage is rooted in the Eastern Cape. She determines the best method of responding to the message is not to work through government channels, but to deliver her assessment directly to the people. This is a hopeful move, one that respects the autochthonous knowledge and agency of a people to save themselves, and which comes in time for the people to identify and pursue a course of action.

Where *Triangulum* suggests the need for a re-balancing, *The Old Drift* presents a nascent vision of that re-balancing. This news is delivered by one of the defining elements of the novel: a classical dramatic chorus comprising mosquitoes. 'Lusaka survived, that dusty plateau, as its own city-state. Kalingalinga became its capital. A small community, egalitarian, humble. People grow all of the food that they eat. There are a few clinics, and one or two schools. Beads are used for barter and voting' (563). Serpell's nearfuture is a return to African values, a ledger rewritten to account for the Earth's capacity and its peoples' dignity. While that return is possible for only some of the former residents of the Zambezi basin, it is an existence that offers sustainability, or at least a pause to catch our breaths in the long, slow swirl of history.

Each of these novels enacts the central elements of Okorafor's definition of Africanfuturism. Each centres African history and points of view. Each is interested in technology and largely concerned with 'what is and can/will be' while carrying 'what has been' (Okorafor). Though Serpell's vision of an egalitarian Lusaka city-state is arguably more optimistic than *Triangulum*'s warning of impending doom, both novels present nearfutures potentially liberated from (neo-/post-) colonial ambivalence. Because black-tech amplifies this ambivalence, liberation, if it is to be accomplished at all, must happen at the site: the use of technologies by the subjects of the (neo-/post-)colony. Reading these texts as examples of black-tech inflected, Okorafor-informed Africanfuturism, we are able to see an eschewing of the technologies of oppression in favour of a re-balanced world shaped by southern African cultures, histories, and points of view.

WORKS CITED

Bhabha, Homi. *The Location of Culture*. London: Routledge, 2004.
—— 'Of Mimicry and Man: The Ambivalence of Colonial Discourse'. *October* Vol. 28, 1984: 125–133.
Bhabha, Homi. 'Notes on Globalisation and Ambivalence'. In *Cultural Politics in a Global Age: Uncertainty, Solidarity and Innovation*, David Held, Henrietta L. Moore, and Kevin Young (eds). Oxford: Oneworld, 2007: 36–47.
Ducournau, Claire. 'The Ambivalent Portrayal of Colonization in the Memoirs of Amadou Hampâté Bâ', *Research in African Literatures* Vol. 46, No. 3, 2015: 68–84.
Fanon, Frantz. *Black Skin, White Masks*. Translated by Richard Philcox. New York: Grove Press (Revised Edition), 2008 [1952].
Haynes, M. 'Black Speculative Fiction Definitions'. www.mhaynes.org/black-speculative-fiction-definitions (accessed 2 June 2020).
Mda, Zakes. *The Heart of Redness*. New York: Picador, 2003.
Mond, Ian. 'Ian Mond Reviews *Triangulum* by Masande Ntshanga'. *Locusmag*, 15 August 2019. www.locusmag.com/2019/08/ian-mond-reviews-triangulum-by-masande-ntshanga (accessed 15 September 2020).
Okorafor, Nnedi. 'Africanfuturism Defined', 15 August 2019. www.nnedi.blogspot.com/2019/10/africanfuturism-defined.htm (accessed 12 September 2020).
Omelsky, Matthew. 'Chris Abani and the Politics of Ambivalence', *Research in African Literatures* Vol. 42, No. 4, 2011: 84–96.
Ndlovu, Siphiwe Gloria. 'South Africa's Dystopian Future Could Easily be Imagined through its Dystopian Past' – Masande Ntshanga chats to Siphiwe Gloria Ndlovu about his new novel, *Triangulum*. *Johannesburg Review of Books*, 3 June 2019. www.johannesburgreviewofbooks.com/2019/06/03/south-africas-dystopian-future-could-easily-be-imagined-through-its-dystopian-past-masande-ntshanga-chats-to-siphiwe-gloria-ndlovu-about-his-new-novel-triangulum (accessed 15 August 2020).
Ntshanga, Masande. *Triangulum*. Columbus, OH: Two Dollar Radio, 2019.
Peires, Jonathan. *The Dead Will Arise: Nongqawuse and the Great Xhosa Cattle-Killing Movement of 1856–7*. Bloomington: Indiana University Press, 1989.
Said, Edward W. *Culture and Imperialism*. New York: Vintage, 1994.
—— *Orientalism*. New York: Vintage, 1979 [1978].
Serpell, Namwali. 'Learning from the Kariba Dam'. *The New York Times Magazine*, 22 July 2020. www.nytimes.com/interactive/2020/07/22/magazine/zambia-kariba-dam.html (accessed 12 Sept. 2020).
—— *The Old Drift*. New York: Hogarth, 2019.
Spivak, Gayatri Chakravorty. *A Critique of Postcolonial Reason: Toward a History of the Vanishing Present*. Cambridge, MA: Harvard University Press, 2003.
Young, Robert J.C. *Colonial Desire: Hybridity in Theory, Culture and Race*. London: Routledge, 1995.

Woman of the Aeroplanes & *the Prediction of the Future*

EZEIYOKE CHUKWUNONSO

INTRODUCTION[1]

Among many writers and critics, there is almost a unifying phrase employed in describing Kojo Laing, that is that he was underappreciated while alive[2] (Obi-Young 'Is He African Literature's Greatest Linguistic Innovator?'; Ryman '100 African Writers of SFF – Part Fifteen'). Geoff Ryman described the news of Laing's death thus: 'The Ghanaian press were curiously silent at first about the death of this great writer'. This chapter suggests that the reason Laing's work was underappreciated while he was alive was due to it being misread as magical realism and thus classified as 'hard to understand'. Since his writing resisted the imposition of a magical realist framework, it was pushed to the margins and relegated to the realm of the obscure. Laing's marginalization in the canon was exemplified by the lukewarm way the academic community reacted to news of his death. This chapter will first show how Laing's work has been misread as magical realism, by using his novel *Woman of the Aeroplanes* (1988) as a case study. It will then employ a 'political futurist' framework in order to analyse the text, centring an improved political economy between global South and global North. 'Political futurism' can either be dystopic or utopic. It positions itself against the scientism and progressivism inherent in some 'technocultures' that are underpinned by capitalist ideology with a naïve faith that progressiveness or superiority in technology will end inequality and environmental problems. 'Political futurism' looks at societal webs or what Arne Johan Vetlesen calls 'exchange and mutual dependency' (*Cosmologies of the Anthropocene Panpsychism, Animism, and the Limits of Posthumanism*: 59). Its dystopia is where this exchange and mutual dependency is fraught, while its utopia presents a society without gender, race or class oppression and of equality and in which the environment is adequately protected. This paper makes the case that 'futurism' in speculative fiction is not exclusive to 'technoculturism' just as dystopia is not only technologically oriented. For example, *1984* by George Orwell (1949) or *The Handmaid's Tale* by Margaret Atwood (1985) are two works of political dystopia in the Western canon. *Woman of the Aeroplanes (1988)* is their opposite as a novel of political utopia.

Mark Bould ('Afrofuturism and the Archive': 13) suggests that most African science fiction (SF) has been misread because it has been read within the paradigm of magical realism. Conceptualizing this, Stephen M. Hart writes:

> magical realism is born, the novel suggests, in the gap between the belief systems of two very different groups of people. What for the inhabitant of the 'First World' is magical … is real and unremarkable for the inhabitant of the 'Third World'. ('Magical Realism: Style and Substance': 4)

Christopher Warnes identifies that the employment of the term 'magical realism' comes from the need by critics to engage with postcolonial fiction through the critical framework of 'writing back'. Warnes writes:

> Generalising accounts abound describing the ways magical realism represents the 'writing back' of the margins to the centre, how it blurs the binaries of modern thought, how it critiques the assumptions of the Enlightenment, how it shows up the limitations of European rationalism, how it reveals the ethical failings of realism. (*Magical Realism and the Postcolonial Novel*: 6)

Warnes's observation of the link between magical realism and 'writing back' becomes clearer when the way magical realism is conceptualized is foregrounded. In magical realism, the speculative function of the literature is understood as realism from the perspective of indigenous cultures, while it is viewed as magical from the perspective of the West. The aim of this manoeuvring is an attempt by critics to straightjacket these speculative works to be read as a reaction against colonialism and its aftermath. By positing that indigenous speculative fiction showcases indigenous epistemologies or modernity, which sits at odds with Western modernity/epistemology, indigenous speculative fiction is fitted into the 'writing-back' paradigm made popular by *The Empire Writes Back: Theory and Practice in Post-Colonial Literature* (1989) by Bill Ashcroft et al. This work, which argues that fiction written by once-colonized people serves as a reaction to colonialism and its aftermath, showcases the modernity/epistemology of the 'other' as a counter claim to the West's justification for colonization in bringing modernity to the 'other'. Magical realism is thus centred in contestations that modernity is not a monolith. As such, magical realism reduces African/indigenous speculative fiction to nothing more than a mode of 'writing back', showcasing an alternative modernity or epistemology to the West. However, this approach, apart from its naïve oversimplification of reducing the complex realities of African/indigenous people and their aspirations portrayed in their fictions, to become just a mere reaction to colonialism, also serves to 'de-science' these SF works, to use Bould's term ('Afrofuturism': 13).

Sofia Samatar highlights that, before this recent boom in African SF that has led people to start recategorizing the text, *Woman of the Aeroplanes* 'circulated simply as African literature or perhaps magical realism' ('Toward a Planetary History of Afrofuturism': 175). For instance, Brenda Cooper reads the novel as magical realism in conjunction with the works of Ben Okri and Syl Cheney-Coker. In her analyses of *Woman of the Aeroplanes*, she posits that the novel is 'increasingly difficult to sustain. The paradox lies in the fact that the space that is being so jealously guarded is the ability to mix and cross races, countries and cultures. Eventually, paradox begins to crack and become contradiction' (*Magical Realism in West African Fiction*: 204). The paradigm of magical realism which Cooper employs to read the novel fails her, but instead of acknowledging this reality, Cooper turns the blame to the novelist for their failure to sustain the paradigm she imposes. This approach in Cooper's analysis is a trend Warnes notices among the critics using magical realism in textual analysis. Warnes notes that among these critics 'texts are used to understand magical realism, rather than magical realism being used as a tool to unpack and interpret texts' (*Magical Realism Realism and the Postcolonial Novel*: 18). This chapter seeks to reposition the novel as a speculative work outside the constraints of magical realism and 'writing back', by reading it as a futuristic text that constructs a utopic trade between global North and global South. This reading takes its cue from Kodwo Eshun's suggestion of the need to have African futurist analyses. Eshun argues that African intellectuals have largely abandoned futurist analyses, while identifying why re-routing towards futurist readings is important ('Further Considerations of Afrofuturism').

IMAGINING AND CONTROLLING THE FUTURE

Eshun notes that imperial racism and the desire to overturn it, that is the desire to write back to it, 'overdetermined Black Atlantic intellectual culture for several centuries' ('Further Considerations' 287). Because of this, among Black Atlantic intellectuals (i.e. intellectuals of African descent), '[f]uturological analysis was looked upon with suspicion, wariness, and hostility' (288). According to Eshun, the result of this, among other things, was that Black Atlantic cultural activists 'little by little, ceased to participate in the process of building futures' (289). There is a grave danger in this. Since, in the current world, information about the future is an 'increasingly important commodity' (290), the 'powerful employ futurists and draw power from the futures they endorse, thereby condemning the disempowered to live in the past' (289). So, in foreclosing any other narrative, apart from that of colonial discourse (the writing-back mandate), Black intellectuals risk condemning the once-colonized to remain perpetually disempowered and entrapped at the mercy of the future which the colonizer has drawn up and endorsed. It is because of this that Eshun notes science fiction's role in 'the prediction and control of tomorrow' (290). This controlling of one's tomorrow is a prerequisite of survival in the contemporary world. That is, futurological analysis is intrinsically important in and for itself, and its validity does not lie in it being a form of protest, but rather in the dimensions of its ontoethics. Elizabeth Grosz explains ontoethics as 'the possibility of the future being otherwise than the present' (*The Incorporeal: Ontology, Ethics, and the Limits of Materialism*: 2); that is, the imaging of better relationship across the humanity than what reality presently offers us.

In *Woman of the Aeroplanes*, Kojo Laing engages with the business of future prediction. Laing provides an alternative vision of Africa's political economy and ethics, which moves away from one overdetermined by late capitalism. In doing so, Laing wants to control the present by remodelling the 'now' into this future projection, akin to Fredric Jameson's suggestion that the function of SF is in 'transforming our own present into the determinate past of something yet to come' (2005:288). According to Jameson, the 'political unconscious – is a construct' and should 'be re-constructed on the basis of empirical 'texts' of all sorts' (*Archaeologies of the Future*: 283) in order for change to occur. *Woman of the Aeroplanes* is a project that aims to bring a political change by engaging in the reconstruction of a political unconscious that has been saturated with capitalist ideology and re-routing it to that rooted in the ethics and political economy of Ujamaa.

The future projection in *Woman of the Aeroplanes* is not ignorant of current historical facts. What it does is resist the present's ability to over-determine the future by imagining outside of the present to construct the future. The present, according to Jameson, has been controlled to produce a future which still mirrors it, in which power structures are unevenly distributed in a combined and uneven development. Jameson describes this process as the way 'SF registers fantasies about the future' (*Archaeologies of the Future*: 286). Alena Rettová articulates this as SF's power to 'project possibilities and model alternative worlds [which] has a different ontological status from the possible' ('Sci-fi and Afrofuturism in the Anglophone Novel': 159). In acknowledging the historical past of the present world, and re-imagining different ontological statuses for it, Laing wrote: 'You wonder why God created them in the first place, and then when you can't get a biological reason for the so-called backwardness then you look for cultural reasons … Of course economic ones can't possibly exist' (189–190). In the logic of late capitalism, cultural difference is foregrounded as the cause of slavery, colonialism, and racism. That is, these injustices happened because of the establishment of binaries that made the West sees the rest as the 'other'. This logic ironically avoids any hint of economic intent in that injustice. Laing sees this logic as a ruse. He then foregrounds the capitalist intent of the conquest, '*Of course economic ones can't possibly exist*' (190, emphasis added), as the driving force for the injustice. Laing paints the theme of *Woman of the Aeroplanes* to be that of political economy, in order to reflect a better business relationship between Africa and the West, or a better 'ontological status', to use Rettová's words. The reimagined relationship offers up a different possibility beyond the current one which is dominated and controlled by capitalist ideology. Laing does this with the hope of modelling the present to become a useful past of the imagined future.

Woman of the Aeroplanes is a story of two towns: Tukwan, in Ghana and Levensvale, in Scotland. They discovered that they are linked together by their immortality and so they set up trading arrangements to cement this special relationship. The aeroplanes referenced in the novel's title, and central to the framing of the protagonist, Pokuaa, serve as a motif for an ideal future trading relationship between the global South and the global North. Laing writes that Pokuaa

> was the mistress of two small aeroplanes … The planes were owned in trust for the town by Pokuaa: she bought them, and had arranged for the town to buy them back by exporting palm-nuts and cassava to a

> sister town in the UK … Now they were waiting patiently for one great thing: to make a journey to the sister town, Levensvale, a journey that they hoped would lead to prosperity. (6)

The first striking thing here is how the means of production is controlled by the town, Tukwan. It is a town which is banished from the rest of Ghana because it is subversive to Ghana's capitalist political economy. Ghana ostracises Tukwan with the hope that it will die a natural death. Pokuaa, their leader, buys the planes as she projects that this will result in a flow of cash into the town. Since the means of production in the town is owned by the public, Pokuaa orchestrates the community's control of their income by enabling the community to gain possession of the planes through a buy-back clause. In this way, the prosperity which flows in from foreign trade can be equitably shared by all and not just enter into a single pocket, a situation which would create division between the poor and the rich. Pokuaa knows, as Nkrumah posits, that 'true economic and social development cannot be promoted without the real socialisation of productive and distributive processes' ('African Socialism Revisited'). It is because the economy of the town is built in this way, with a fair distribution of wealth, that the sense of community among its inhabitants becomes clearly palpable: 'Now they were waiting patiently for one great thing: to make a journey to the sister town' (6). There is this great sense of the collective in this passage, in the way the community sees and imagines themselves. It is on this sense of a utopian Ujamaa economic blueprint that Julius Nyerere anchored the following statement: 'we were individuals within a community. We took care of the community, and the community took care of us' ('Ujaama – The Basis of African Socialism': 7). It is this ethical familyhood of Ujamaa which leads Pokuaa to freely surrender her gift; her ingenuity in trading in service of her community. The planes are not extracted from her and put into use by the community through the brutal force of the state, as might be the case in a totalitarian Marxist regime; rather, she gives up her claim of ownership out of her own free volition. In doing so, Pokuaa serves as the leader of a society built on the ethics of extended family and Ujamaa. It is not just she who lives by these ethical political guidelines.

The town is made up of people who have their own individual gifts and who willingly surrender them in service of the common good. For instance, there is Kwame Atta, the bad twin, and a renowned inventor. The community depends on his scientific inventions to survive. He knows he could make vast amounts of money from them if he leaves the town and travels to Accra or Kumasi, but he chooses to remain in

Tukwan not because of constraints but because of the political ethics of Ujamaa ruling the town, of which he believes in. His father, Mackie, detests this decision:

> He could easily have made more money by escaping from this crazy time here, and going to Accra or even Kumasi ... but no! He makes a fool of himself by inventing all sorts of fantastic bolastic things that give him a history but no status, no money yet. (158)

Apart from Kwame's father who is depicted as a lazy man, the rest of the citizens are hardworking, with a range of talents which would make them rich if they escaped the town. There is the talented lawyer, Tay; Maimuna, the accomplished hotel manager; they also have the brilliant Pastor Mensah, but they all prefer to stay instead of commodifying themselves in a capitalist world-system where they would have no control over their jobs. In doing this, Laing projects a vision of the world that subverts the view of capitalism as the only viable political economy, with other systems doomed to collapse because they go against human instinct. As capitalist systems counter the natural human inclination of Ujamaa political economy-cum-ethics, capitalism is artificial to wo/man. It has built its ideological construct in a way that Immanuel Wallerstein identifies as punishing those who do not accumulate capital while rewarding those who do ('The Modern World-System and Evolution': 4). So, even when many are averse to capitalism because it is anti-human in its ethics, the fear of being punished for not accumulating capital and the harsh consequences of this punishment drives everyone to grotesque commodification.

Since Tukwan is built on alternate political economy and ethics to capitalism, the trade which Pokuaa initiates with their sister town, Levensvale, in Scotland, becomes a trade which arises primarily from the need of both towns. It is not the sort of a trade in which one party is structurally forced to respond to the needs of the other party, as per the contemporary division of the world market between the global South and the global North. Nor does it neglect the needs of its citizens as in the case of countries like India, who while they are a big exporter of rice, contain within their borders some citizens who starve from malnutrition (Young, *Postcolonialism: A Very Short Introduction*: 135). As the people of Tukwan wait with excitement for the future of their trading relationship, a similar ripple of hope is seen among the people of Levensvale, who believe that 'Mackie, business man plenipotentiary ... had arranged a good business deal with Pokuaa of Tukwan over aeroplanes, cassava and palm-nuts' (63). Seeing this as good business, the town of Levensvale awaits the coming of representatives from Tukwan with celebration and openness.

To have this fair trade, which benefits both parties, Laing first builds factors to make it possible. Notwithstanding that one town is 'Black' whereas the other is 'White', Laing connects them in a magical relationship in which people from both towns are immortal. This immortality makes the two towns, though continentally different, racially opposite, and culturally dissimilar, still able to link and relate to each to a greater extent than to surrounding cities on the same continent. Immortality serves as a figuration of ideology which the two cities share, whereas their close neighbours uphold different ideals. It is through this shared ideology that both towns are able to transcend their racial differences. Pastor Korner Mensah declares in a heated debate about the history of racism that, among them, 'our two towns must give each other the greatest thanks for having none of this terrible nonsense with us, none of this horror, none of this assault on the colour of the soul' (190). Here, the utopia which Laing imagines, which is informed by that of Sartre, is the emergence of a classless, 'colourless' society devoid of any form of oppression based on gender, race, or social status.

Arriving at this utopic point between the two towns does not happen in rejection of history, or in claiming amnesia when it comes to historical facts; the towns instead are conscious of what they have chosen to become. One can see this, for instance, in the joint business venture of the two towns. The people of Levensvale agree to pay reparations for the historical oppression of their sister town's ancestors, knowing full well that it was a crime that emerged from capitalism, with all its inherent immorality, and that restitution should at least be made in economic terms. The money they pay Tukwan is not marked as aid, nor is it classified as charity or a loan, but instead as restoration. On the other hand, the people of Tukwan consciously do not try to view their history as starting from the historical injustice of racial atrocity towards them. That is, the timeline of their history is not structured under the precolonial-colonial-postcolonial paradigm: 'History never walks here, it runs in any direction. We have been building something different here for years and years' (1). In Africanfuturism, time that runs in any direction becomes what Grosz calls a 'virtual selection' (*The Nick of Time: Politics, Evolution, and the Untimely*: 258) in which neither the present nor the past is privileged to determine the future. The future is a conscious selection from the present and the past which serves as raw data to inaugurate an eruption, a radical futurism different from both. This futurism differs from Bhabha's hybridity where the past and the present meet in an antagonistic confrontation. Futurism in hybridity privileges the present as it blindly assimilates and syncretises the past into the present. It thus becomes a futurism of a hybridized identity; a futurism of coping and adapting to (post) colonial trauma. History as it unfolds implies a movement from point A to B, from colonialism to postcolonialism. This is the trap for every Ghanaian city surrounding Tukwan as, since their history is a linear progression, reflected in the formation of a story with a beginning, middle, and end, the capitalist world-system has totally imagined and contained them. This is why science fiction, as Jameson noted, refuses an ending, and resists being totally contained.

The city of Tukwan is able to make progress and build something different, because they resist being contained through opting for a history that can run in any direction. The advantage of resisting this linear and dichotomizing history, split into different antagonist segments (with modernism claiming to be superior to premodern, and postmodern to modern), is that the town is able to exist in harmony with itself, and thus able to progress using all the knowledge resources of their forbearers, along with ones borrowed from outside sources, to make something new. This is what Dilman Dila refers to when he said in an interview: 'No country can 'develop' if they do not see science and technology in their own past', while adding, 'I don't think local innovations can work as islands separated from imported ones. I see a merging of the various knowledge systems' ('An Interview with Dilman Dila': 30). Tukwan, in comparison to the rest of the cities in Ghana, has the advantage of viewing science as a part of them while borrowing from other sources of knowledge in order to progress. Their approach also reveals what is lacking in the imposition of linear readings of the history of precolonial-colonial-postcolonial periods, in which everything before colonization is framed as pre-historic, with modernity hinging on contact with the colonizer. In *Woman of the Aeroplanes* we see the cities that surround Tukwan embrace this approach and in turn, dwell in poverty. In a sense, modernity becomes foreign and rooted in an epistemology in which the West who brought modernity, holds a patent to it. The West then sets the terms and the conditions of this procurement, creating a situation which establishes a perpetual dependence.

An important factor, which paves the way for fair trade between the two racially/culturally different towns, is that the two cities hold each other as equals and as comrades. We see this, for instance, when delegates from Tukwan depart for Levensvale for a business summit. Before this important journey, there is thorough preparation, with meticulous selection applied for those who are going to represent the town. So detailed is the assessment that even non-humans are considered in it. Laing writes: 'But it was a sign that the forest too must be visited, must be assessed against the savanna … But there was a wild flight when it was revealed to the lake that only two ducks would travel, plus one elephant grass and one crabgrass uprooted' (55). Every decision made in this assessment, from human to non-human, is based on what will serve the common good of all in the community.

Notwithstanding the time and energy afforded to their preparations for this journey, the people of Tukwan do not perform the basic requirements guiding travel in capitalist world economies, which is the adequate application of immigration documents for all the delegates. This is not an oversight or forgetfulness; it is that the sense of equality between them and their sister town means that immigration processes are not an important need requiring attention. A similar air of reasoning underlines those in Levensvale too. But this principle of equality which is enjoyed by the two towns is undoubtedly not shared by those living within the capitalist world economy, where inequality is a necessary part of the system, driving its ethics. Thus, every human interaction in this capitalist system is an opportunity for exercising inequality, to remind and make the other feel lesser and inadequate. In eliminating inequality, the whole capitalist world economy crumbles. The need to perpetuate these ethics of inequality, with the rest of the UK not sharing in Levensvale's ideology, drives immigration officers to confront the delegates from Tukwan when they arrive in Levensvale. The immigration officers come from the town outside of Levensvale because, although autonomous in certain things, they are still subject to broader controls when it comes to some administrative issues. Laing paints the moment thus:

> Immigration John spoke at last … 'We believe, not all of the party has visas, we believe…' At last Mackie said, with a growl, 'I would wish to contact the Lord Provost'. 'Politics won't get you anywhere in this case, for we have taken the opportunity to warn the Provost of the political consequences of unorthodox measures in immigration matters such as this. (79–80)

The matter is later resolved with the help of the Lord Provost, who obtains visas for both humans and non-humans from Tukwan. However, what Immigration John achieves in this interaction is announcing to the delegates what their race, status, and class are. That is the obsession of the capitalist unconsciousness, to always differentiate in order to repeatedly announce 'unequalness'. With the immigration officer gone, the delegates from Tukwan and the citizens of Levensvale slip back to a life of mutual respect and equality. Laing writes:

> The Tukwan law of genetics was accepted in Levensvale, even by the Lord Provost ... now since a snore in Levensvale could originate in Tukwan ... everybody was free to be and do what he or she liked. There was a blast of freedom from freely-mixed bodies and worlds, ampa'. (86)

This equality does not demand that a group of one people be assimilated into the culture of the other before it is accorded. The two cultures still differ from each other, ranging from the type of food (87), clothing (76), and also the way they practise their religions (24, 190), but the immortality they share, which is figurate to the ideology of socialism, allows them to transcend these differences.

Furthermore, that the two towns are able to gain from fair trade to the benefit of both boils down also to the independence they enjoy from each other. This independence is not just territorial but, most importantly, it is also economic. On their way to Levensvale, missing home, the character of Pastor Mensah laments: 'What sort of nonsense soul is this new town that we have set up ourselves through our own engineered ancestry? What do we want with money or with ideas that we already have enough of' (61). From Pastor Mensah's speech, it can be inferred that, even though the journey is undertaken with the hopes of making the people of Tukwan more prosperous, they are inwardly economically independent in their own right and so are in a better position to bargain for a fair trade. They are at home with their culture and identity and are able to build as a subversively progressive city after being banished from the rest of Ghana. It is this independence, being both in the world and also out of it, that makes them re-think their political economy, and they are able to adopt a Ujamaa political economy and ethical framework. In explaining the gain accrued for self-sufficiency in independence, Victor Li, uses the metaphor of isolation in the film *Cast Away* to explain the point thus: 'it is precisely the project of self-reformation and self-care, made possible by Noland's isolation from the demands of globalization, that also enables him to re-enter the global economy as a better, more valuable citizen-worker' ('Globalization's Robinsonade': 63). In a similar way, Laing imagines the utopia of Tukwan as being able to become stronger, with a better economy, because it first moved to self-care by isolating itself from the world and is then able to re-enter the global economy under better terms. Laing knew that, when the developed world pushed for the increased opening up of 'third-world' territories, it was with the intent of assimilating and hybridizing them, thus making them perpetually dependent on the 'core states'. In that way, any trade entered into with the core state will be unequal and unfair, since they determine the tone of the trade. Under current World Bank and International Monetary Fund (IMF) structures, the core states regulate and control trade with the developing nations through exchange rates and loans, to name just two examples. Laing projects a future in which the third-world economy has severed ties with these institutions, isolating themselves from the pull of the 'first world' in order to build for themselves. They later collaborate on equal terms and conditions with other peoples. This collaboration will not be based on a similarity of racial identity but rather on a shared ideology of an economy built for the many and not for the few. That is, Laing is offering us a better model of globalization outside the present asymmetrical one marred with exploitation and unevenness.

CONCLUSION

There is of course a genuine reason why Laing is misread. Surrounded by texts which Fredric Jameson describes as an 'allegory of nationhood' (*Archaeologies of the Future*: 69), the canon that begun to be formed from the 1950s centred postcolonial literature that engaged in 'writing back'. Laing's writings were distinct from those of his contemporaries and were far ahead of their time. Misreading occurred, but not because Laing's work is speculative fiction. African writing has always been speculative, even before the emergence of the 'writing-back' paradigm, which invariably swept speculative writings under the carpet from the mid-twentieth century. Take for instance the tradition of which *Shaka* (1910) by Thomas Mofolo or D.O. Fagunwa's *The Forest of a Thousand Daemons: A Hunter's Saga* (1938) are part.

For Laing, the type of decolonization of 'writing back' posed a number of the pitfalls, of which Jameson says:

> Overhastily, I will suggest that 'cultural revolution' as it is projected in such works turns on the phenomenon of what Gramsci called 'subalternity,' namely the feelings of mental inferiority and habits of subservience and obedience which necessarily and structurally develop in situations of domination – most dramatically in the experience of colonized peoples. (*Archaeologies of the Future*: 76)

What Jameson shows is that subversion inherently sustains a master-slave relationship and so should be a careful tool to deploy. A cultural heritage that is permanently built on it damages more than it solves. What this process secured then is the agency of the history as described by the colonizer. This implies that it is the colonizer that retains control over the colonized in their determination as whether to handover a token of rights or to withhold it, having listened to the protests of works that 'write back'. Laing engages in political futurism that bypasses this approach, by proffering a form of decolonization that exists beyond this binary; a decolonization without a master-slave relationship in which the enslaved person is protesting to their master. It is now left for us to make the choice of which model of decolonization to employ. But whatever choice we make, we ought to be aware that the 'writing-back' paradigm does more disservice to African cultural and economic growth by locating the continent forever in the past, while the present and future remain limited without alternative 'becomings' and eruptions of newness. Colonialism in this way becomes forever present, with an unending future.

NOTES

1 This chapter is an extract of my ongoing PhD. thesis. So, I am grateful to my supervisory team especially to my Director of Studies, Dr Chloe Germaine Buckley for her feedback and support.

2 A more detailed overview of this sentiment is also in the tributes by different writers that poured in after Laing's death. https://johannesburgreviewofbooks.com/2017/05/01/rest-in-power-pa-kojo-paying-tribute-to-kojo-laing.

WORKS CITED

Ashcroft, B., Griffiths, G., and Tiffin H. *The Empire Writes Back: Theory and Practice in Post-Colonial Literatures*. London: Routledge, 2002.

Bould, M. 'African Science Fiction 101'. *SFRA Review* No. 311, 2015: 11–18.

Cooper, B. *Magical Realism in West African Fiction: Seeing with a Third Eye*. London: Routledge, 2004.

Dila, D. with Polina Levontin. 'An Interview with Dilman Dila'. Special edition: African and Afrodiasporic SF, *Vector* No. 289, Summer 2019: 24–30.

Eshun, K. 'Further Considerations of Afrofuturism'. *CR: The New Centennial*

Review Vol. 3, No. 2, 2003: 287–302.
Grosz, E. *The Incorporeal: Ontology, Ethics, and the Limits of Materialism*. New York: Columbia University Press, 2017.
—— *The Nick of Time: Politics, Evolution, and the Untimely*. Crows Nest, NSW: Allen & Unwin, 2004.
Hart, M.S. 'Magical Realism: Style and Substance'. In *Companion To Magical Realism*, M.S. Hart and O. Wen-chin (eds). Woodbridge: Tamesis, 2005.
Jameson, F. *Archaeologies of the Future: The Desire Called Utopia and Other Science Fictions*. London: Verso, 2005.
—— 'Third-World Literature in the Era of Multinational Capitalism'. *Social Text* No. 15, Autumn 1986: 65–88.
Laing, K. *Woman of the Aeroplanes*. London: Heinemann, 1988.
Li, V. 'Globalization's Robinsonade: "Cast Away" and Neo-Liberal Subject Formation'. In *Rerouting the Postcolonial: New Directions for the New Millennium*, J. Wilson, C. Sandru, and L.S. Welsh. London: Routledge, 2010: 60–71.
Malec, J. 'Rest in Power, Pa Kojo: Paying Tribute to Kojo Laing'. *Johannesburg Review of Books*, 1 May 2017. https://johannesburgreviewofbooks.com/2017/05/01/rest-in-power-pa-kojo-paying-tribute-to-kojo-laing (accessed 20 January 2021).
Nkrumah, K. 'African Socialism Revisited'. 1967. www.marxists.org/subject/africa/nkrumah/1967/african-socialism-revisited.htm (accessed 10 September 2020).
Nyerere, K.J. 'Ujamaa – The Basis of African Socialism'. *The Journal of Pan African Studies* Vol. 1, No. 1, 1987: 4–11.
Obi-Young, O. 'Is He African Literature's Greatest Linguistic Innovator? The 5 Books of the Late Kojo Laing'. *Brittle Paper*, 10 May 2017. https://brittlepaper.com/2017/05/africas-novelist-5-books-late-kojo-laing (accessed 5 January 2021).
Rettová, A. 'Sci-fi and Afrofuturism in the Afrophone Novel: Writing the Future and the Possible in Swahili and in Shona'. *Research in African Literatures* Vol. 48, No. 1, 2017: 158–182.
Ryman, G. '100 African Writers of SFF – Part Fifteen: Ghana Introduction: A Golden Coast'. *Strange Horizons*, 2020. http://strangehorizons.com/non-fiction/introduction-a-golden-coast (accessed 20 August 2020).
Samatar, S. 'Toward a Planetary History of Afrofuturism'. *Research in African Literatures* Vol. 48, No. 4, 2017: 175–191.
Vetlesen, A.J. *Cosmologies of the Anthropocene Panpsychism, Animism, and the Limits of Posthumanism*. London: Routledge, 2019.
Wallerstein, I. 'The Modern World-System and Evolution'. *Journal of World-Systems Research*: Vol. 1, No. 19, 1995: 1–15.
Warnes. C. *Magical Realism and the Postcolonial Novel: Between Faith and Irreverence*. New York: Palgrave, 2009.
Young, J.C.R. *Postcolonialism: A Very Short Introduction*. Oxford: Oxford University Press, 2003.

Re-membering *the Past*

Black Panther, Sovereignty, & the Cultural Politics of Africanfuturism

KAYODE ODUMBONI

INTRODUCTION

Ryan Coogler's *Black Panther* elicited a unique intellectual and cultural excitement among its audience – from the critical audience to the layman viewer – when it was released in 2018. The movie's wide appeal cannot be divorced from its adept deployment of the genre of speculative fiction, a genre whose audience is increasingly growing worldwide. And to people for whom science fiction or superhero movies are not a go-to, *Black Panther* still holds its appealing power because of the compelling message it conveys about the condition of global Blackness. As a movie that articulates an unconventional vision of what it means to be Black, *Black Panther* successfully alters – in a radical way, really – the ways in which the ontological temporality of Black life is often contemplated in contemporary discourses. In other words, the movie initiates a drastic shift in the way we think about the past, the present, and the future of Africa – where Africa stands as a geographical metonymy for Blackness.

With a growing body of scholarly criticism centred on *Black Panther*, the movie has been variously analysed as a work of science fiction that imagines an alternative past for Africa in order to envision a radically different (almost an other-worldly, impossible-to-achieve-in-reality) future for Africa and the people of African descent (Asante and Pindi '(Re) imagining African Futures: Wakanda and the Politics of Transnational Blackness': 220). In this chapter, I explore the ways in which *Black Panther* maximizes the many aesthetic affordances of speculative fiction to re-capture a past that is not merely alternative history. In my reading, the movie actually *re-presents* a true-to-life past in order to jolt the reader's memory and to disrupt the hegemonic historiography that has conditioned how Africa's historicity is understood. Whereas *Black Panther*'s portrayal of Wakanda as an African nation that was never colonized has been commonly interpreted as the movie's engagement with some 'mythical memory of the past' – a reading that often leads to critics arguing that the film 'lacks an engagement with African historical specificity' (Asante and Pindi: 222) – I argue that *Black Panther* actually depicts a historically accurate past and, even more so, it does that with the political imperative of *re-membering* a history that has been *dismembered* by the epistemic violence engendered by global White supremacy. Approaching *Black Panther* with this understanding, that the historical past portrayed in the movie may have deployed fantasy but is definitely not mythical, allows for a proper appreciation of the movie's actual aesthetic and political vision.

In this chapter, I unpack the various ways in which *Black Panther* employs the elements of speculative fiction and signifies upon the register of Africanfuturism specifically in its recuperation of Africa's history. Also, I situate my use of the term Africanfuturism within the long, treacherous history of both speculative fiction and Afrofuturism, and establish why I read *Black Panther* as a work of Africanfuturism, as opposed to other genre labels. Ultimately, my analysis of the movie explicates the postcolonial impulse with which the movie grapples with the questions of hegemony and sovereignty in its apprehension of Africa's present, while centring Blackness and charting a course for potential Black futurities.

'HOW FAR BACK SHOULD MEMORY REACH?' *BLACK PANTHER*, COLONIALISM, AND AFRICA'S BELEAGUERED SOVEREIGNTY

In his essay 'Memory, Truth and Healing', Wole Soyinka ponders:

> How far back should memory reach? How deeply into the recesses of the past? The answer that springs spontaneously to mind is that memory is not governed by the statute of limitations, and that collective memory especially is the very warp and weft of the tapestry of history that makes up society. Unravel and jettison a thread from that tapestry and society itself may become undone at the seams. (21)

Much of modern global historiography has been constructed such that it is almost impossible to understand Blackness (or Africa) outside the historical contexts of slavery and colonialism. Nevertheless, the fact remains that slavery and colonialism are not the only historical realities of Africa nor (should be) the only paradigms for understanding Africa and Blackness. But because hegemony and history are inextricably interwoven, the histories of slavery and colonialism, which paradoxically implicate and subjectivize the West, have become the master narratives through which we contemplate Africa. As a cultural production, *Black Panther* dislodges these 'statutes of limitations' that have confined the memory of Africa to 1619 and after. Of course, there was life in Africa before the European incursion. In fact, scholars such as the historian Ade Ajayi have countered the argument that the impacts of European colonialism on Africa are irreversible. According to Ali Mazrui and Michael Tidy, scholars who align with Ajayi's position argue that 'the few decades of European rule in Africa were just a brief interlude when examined against the millennia of Africa's history … we exaggerate the importance of European colonialism because we are so close to it in time' (*Nationalism and the New States in Africa*: xi).

The narrator in the *Black Panther* tells us about 'millions of years ago', signalling from the get-go that the movie invites us to contemplate Africa without taking recourse to the fatalistic logic of its colonial past, to access in depths 'the recesses of the past'. Therefore, the question that *Black Panther* seeks to answer is this: what happens if we think of Africa outside of its colonial history? The answer, of course, is not a simple one. It is as complex in temporality as it is in spatiality; and it is these spatio-temporal complexities that *Black Panther* grapples with. Moreover, *Black Panther* equally attempts to refocus the ways in which we remember the history of colonization itself. Because of the pervasive hegemony of the West, colonial modernity has had an ambivalent halo to it – a dynamic that engenders discourses that attempt to articulate and emphasize what is seen by some as the positive outcomes of colonialism in colonized spaces, such as the arrival of Western civilization, the construction of road networks, infrastructural developments, etc.[1] To contest this colonial history, *Black Panther* explicitly points out that the colonial project was a venture of exploitation.

A crucial aspect of the conflict in *Black Panther*, much like the ontological conflict that has constituted the touchstone of Africa's history, is driven by the question of sovereignty. The sovereignty of Wakanda coalesces around Vibranium, the meteorite that serves as the power(base) of the state. And the contention in Wakanda stems from the external aggression that Wakanda experiences in Ulysses Klaue's – by extension the CIA's and, by further extension, the West's – attempts to get hold of, and dominate the use of, Vibranium. This threat of external aggression against Wakanda resulting from the power tussle for Vibranium is an unambiguous parable that re-enacts the bald facts of Africa's colonial history. The way Ulysses Klaue is attracted to Vibranium is the same way Europe was drawn to the natural and human resources of Africa for exploitation. In *The Wretched of the Earth*, Frantz Fanon alludes to 'the administrative and managerial services [of Europe] whose function it was to discover the wealth of the [colonized] country, to extract it and to send it off to the mother countries' (102). Further, he asserts that 'in a very concrete way Europe has stuffed herself inordinately with the gold and raw materials of the colonial countries' (102). Vibranium is, on the one hand, the metallic resource that powers much of the life and technoculture in Wakanda, an element of *fantasy* – 'fantasy' in the sense of the fantastic, being an extraterrestrial element, but also 'fantasy' in the sense that it is the object that tickles the fancy of the West. On the other hand, Vibranium is a symbolic representation of the rich natural and human resources of Africa that appealed to and fueled the exploitative instincts of Europe. Based on the foregoing, therefore, it is clear that *Black Panther* engages with the history of colonialism, even if allegorically. The overarching plot of the movie – as well as its explicit references to colonialism such as when Shuri calls Agent Ross 'Colonizer' – discounts readings that have implied that *Black Panther* is an ahistorical cultural production for portraying an uncolonized African nation.

Like Africa, like Wakanda. The very present threat that the West poses against Wakanda through the pursuit of Vibranium clearly alludes to the imperialist impulse that animated the colonial project. Klaue's invasion of Wakanda and successful stealing of Vibranium from there threatens Wakanda's sense of sovereignty – what Achille Mbembe calls a 'society's capacity for self-creation through recourse to institutions inspired by specific social and imaginary significations' ('Necropolitics': 13) – which essentially captures what happened during colonialism. Because of Klaue, Wakanda realizes that its autonomy is not as secure as had been imagined, and that there are indeed limits to how much the nation can 'hide in plain sight'. The fact that not even Wakanda's *fantastic* ability to 'hide in plain sight' can keep it safe from the prying eyes of the West speaks to not only the far-reaching impacts of colonialism, but also the ubiquity of modern world-systems such as global capitalism and neoliberalism that worked hand-in-hand with colonial modernity. Seeing as Wakanda can no longer hide in plain sight from other nations of the world, it becomes imperative for the nation to insert itself into global (geo)politics. Today, Africa also does not have the luxury to 'hide' from the nefarious impacts of colonialism, White supremacy, capitalist modernity, neoliberalism, and other such manifestations of Western hegemony that have been constructed to keep the continent subjugated.

In addition to the threats of external aggression against Wakanda's sovereignty, there is a deep-seated internal conflict among Wakandans themselves regarding whether the nation should continue on its path of isolationism or change course and be more embracing of the outside world. This existential national conflict is intergenerational; we see it between King T'Chakka and his brother Prince Njobu in the beginning sequences of the movie, and later in their children King T'Challa and his cousin Erik Stevenson (Killmonger). While T'Chakka and his son insist on Wakanda's isolation, principally as a measure of preserving a certain puritanical Wakandanness that is constantly referred to as 'Our Way', Njobu and his son believe that Wakanda needs to intervene in the affairs outside of its borders. This internal conflict speaks to the fractures that have always marked Black identity, especially as premeditated by the histories of slavery and colonialism, between continental Africa and its many diasporas.

To a large extent, therefore, *Black Panther* reflects on the complications that surround the relationship between Africa and its diasporas. Africa as a metonym for Blackness is oftentimes fractured because the past seems to be refracted differently, depending on the side of the Atlantic where Blackness is located; that is, slavery as the defining past of the diaspora and colonialism as the defining past of continental Africa. By emphasizing the interconnectedness of global Black identity, however, the movie demonstrates that a revisionist apprehension of the past is crucial in charting a pathway into the future.

Apart from *Black Panther's* aesthetic political goal of *re-membering* a dismembered African history, the movie equally offers an intricate meditation on Africa's present. Wakanda is the fictional African country through which Black viewers are able to put into proper perspective where the rain began to beat them, in order to understand better where they are and how they got there. In many ways, Wakanda stands as a synecdoche of the African continent, not in a simplistic way that monolithizes the expansive continent, but mostly for the creative purpose of symbolically capturing the (hi)story of the continent. One question that persistently nags Wakanda is that of sovereignty: how does the state negotiate its sovereignty within itself and in relations with other countries of the world? In that regard, the movie uses Wakanda symbolically to meditate on Africa's *present* (as in temporal existence) as well as its *presence* (as in spatial existence) in the world today.

Because Wakanda's sovereignty – that is, the nation's power to rule itself – coalesces around Vibranium, access to Vibranium determines Wakanda's sovereignty and, in turn, Wakanda's sovereignty determines its exclusive access to Vibranium. In a similar vein, at the centre of the African crisis today is the question of access to resources. In an age of neocolonialism where the global North continues to wield a parasitic control on the economic resources of Africa, the continent's sovereignty is under perpetual attack. In a way, therefore, colonialism is not merely Africa's history; the exploitations and imperialism that attended colonialism are still very much the present-day realities of many African countries.

BETWEEN AFROFUTURISM AND AFRICANFUTURISM: TOWARDS THE DECOLONIZING OF SPECULATIVE FICTION

Science fiction, fantasy, and other genres of speculative fiction have a long notorious history of problematically portraying the violence of colonialism and the brutal conquests of imperialism as heroic endeavors (Burnett 'The Great Change and the Great Book': 134; Langer, *Postcolonialism and Science Fiction*: 1; Eatough 'African Science Fiction and the Planning Imagination': 242). As Joshua Burnett notes, 'speculative fiction remained for so long a purportedly White (and straight, and male, and middle-class) genre and one associated with various forms of racism and colonialism' even though it would later become 'a launching pad for counterhegemonic discourse' maximized by postcolonial writers (136). The Eurocentric epistemology that undergirds speculative fiction is what Afrofuturism sought to redress. But, even so, many critics have pointed out how Eurocentric the idea of Afrofuturism itself is, not merely because the term was coined by a White American critic but also because in many ways its conceptualization still very much centres the West (and the mechanisms of its discursive and cultural hegemony) when capturing Black experience – hence, the rallying call for a new term, especially among Black cultural producers in continental Africa who feel that the aesthetic and philosophical underpinnings of their works are not captured under the register of Afrofuturism (Okorafor 'Africanfuturism Defined'; Wabuke 'Afrofuturism, Africanfuturism, and the Language of Black Speculative Literature').

Mark Dery, who is credited with coining the term Afrofuturism, asks: 'Can a community whose past has been deliberately rubbed out, and whose energies have subsequently been consumed by the search for legible traces of its history, imagine possible futures?' ('Black to the Future': 179). In 'Black to the Future: Interviews with Samuel Delany,

Greg Tate, and Tricia Rose', Dery essentially defines Afrofuturism as 'speculative fiction that treats African-American themes and addresses African-American concerns in the context of twentieth-century technoculture – and, more generally, African-American signification that appropriates images of technology and a prosthetically enhanced future' (180). Underlying Dery's definition and earlier rhetorical question is the constraining of Black history as subsumed under White agency – that is, there is no possibility of thinking about Blackness without how Whiteness has indelibly acted and impacted upon its history, its ontology and, by extension, its future. As Hope Wabuke insightfully captures it, Dery's conceptualization 'lacks room to conceive of Blackness outside of the Black American diaspora or a Blackness independent from any relationship to whiteness, erasing the long history of Blackness that existed before the centuries of violent oppression by whiteness'.

Given the conceptual contestations that have shrouded Afrofuturism,[2] some critics, such as Nnedi Okorafor, have embraced alternative conceptualizations to define their art. One of such terms is Africanfuturism. To Okorafor, who coined the term,[3] Africanfuturism differs from Afrofuturism because the 'default [of Africanfuturism] is non-western; its default/center is African'. Put differently, Africanfuturism is 'similar to "Afrofuturism"' but more deeply 'rooted in African culture, history, mythology and point-of-view as it then branches into the Black diaspora, and it does not privilege or center the West' ('Africanfuturism Defined'). Other scholars like Ytasha Womack (*Afrofuturism: The World of Black Sci-Fi and Fantasy Culture*) and Sofia Samatar ('Toward a Planetary History of Afrofuturism'), however, have emphasized the globality (or 'planetarity', to use Samatar's term) of Afrofuturism and, by effect, have argued that Afrofuturism does not exclude black authors and artists from continental Africa. To these scholars, Afrofuturism emphasizes 'blackness, rather than nationhood', with its focus mostly being oriented 'toward outer space, in which Earth figures as one star among others rather than a map carved up by others' (Samatar: 176). Ingrid LaFleur, quoted in Womack, defines Afrofuturism as 'a way of imagining possible future through a black cultural lens … a way to encourage experimentation, reimagine identities, and activate liberation' (Womack: 9). Hence, as far as these scholars are concerned, it is a misreading to see Afrofuturism as 'confined to a narrow geographical or historical track'; instead, Afrofuturism is a capacious philosophical register that 'neither excludes African science fiction nor ignores its specific contributions' (Samatar: 177).

In my reading, the ultimate distinction between Afrofuturism and Africanfuturism boils down to the latter's explicit politics of postcolonial critique. As Jessica Langer has pointed out, for a very long time, 'where there [could have been] postcolonial science fiction, there [was] instead a real or perceived silence' (*Postcolonialism and Science Fiction*: 1). To my mind, it is this void that Africanfuturism seeks to fill. Africanfuturism engages speculative fiction with an unambivalent aim of disrupting the epistemological structures upon which the history of colonialism and the present realities of neocolonialism subsist. Therefore, Africanfuturism offers a scathing postcolonial, counterhegemonic critique of global modernity. Burnett concludes that 'critics inside and outside the academy are increasingly recognizing the significance of black and postcolonial speculative fiction as discrete genres'. (136) – 'postcolonial speculative fiction' being another name for what Okorafor terms Africanfuturism. As well, Langer charges that

> writers, film-makers and others involved in the production of postcolonial science fiction *participate* uniquely in this process of decolonization, utilizing the particular strengths and possibilities contained in the science fiction genre to further the project of a world not only politically but (variously) economically, culturally, intellectually and/or creatively decolonized. (8)

ENVISIONING THE FUTURE THROUGH BLACK SPECULATIVE FICTION

Black Panther narrativizes a vision of recuperating an African past that had been epistemologically brutalized by the machinations of colonial modernity. Of course, this historical recuperation is crucial because the future (and how to proceed into it) is at stake. Eatough, quoting Alvin Toffler, has pointed out that science fiction is 'a kind of sociology of the future', one whose primary virtue is not in its literary function but in its 'value as a mind-stretching force for the creation of the habit of anticipation' ('African Science Fiction and the Planning Imagination': 243). *Black Panther*'s aesthetic vision, in my reading, is less of 'creating' a habit of anticipation – for, as it stands, global modernity had already conditioned the way Africa's future was contemplated and anticipated before now. Rather, what the movie does is to 'shift' that already engrained habit of anticipation, to radically disrupt the ways in which we imagine Africa's future, by depicting a future of high technoculture in which Africa is contributing significantly – if not being the sole contributor – to global capital and rendering aid to the West, beginning with the United States of America. In her book *Afrofuturism: The World of Black Sci-Fi and Fantasy Culture*, Ytasha Womack notes that the fate of Black characters in sci-fi movies from the 1950s through the 1990s was to be always cast, by default, as 'dour'. 'The Black characters in those films popped up as the silent, mystical type or maybe a scary witch doctor, but it was fairly clear that in the artistic renderings of the future by pop culture standards, people of color weren't factors at all' (Womack: 7). *Black Panther* is consciously speaking back to this tradition of sci-fi movies by contesting the stuck portrait of Black people as objects rather than subjects who can participate in the project of the future.

Black Panther's political and cultural resistance to these conventional portraits of the future vis-à-vis the place of Black subjectivity in science fiction manifests in two ways. Firstly, *Black Panther* contests science fiction's speculation of the future by portraying Wakanda as the technological powerbase of the world, not only in the present but also in the future. As Alondra Nelson notes, it is conventional for 'racial identity, and Blackness in particular, [to be cast as] the anti-avatar of digital life. Blackness gets constructed as always oppositional to technologically driven chronicles of progress' ('Afrofuturism': 1). Wakanda renegotiates such constructions of the future and how Blackness interfaces with that future by portraying Wakanda as the centre of technological advancements. Second, the movie's projection of the future is animated by a race-conscious impulse. The future of Wakanda is neither post-racial nor a-racial. Much of traditional science fiction projects an a-racial future, speculating 'forecasts of a utopian (to some) race-free future and pronouncements of the dystopian digital divide [as] the predominant discourses of Blackness and technology in the public sphere' (Nelson: 1). The vision of the future that we see in the universe of *Black Panther* is a far-cry from such a 'placeless, raceless, bodiless near future enabled [*only*] by technological progress' (1, emphasis added). The future in *Black Panther* does not discount the place of race, nor does it diminish geographical situatedness. Wakanda is still very much 'an African country' in the future, and the geographical rootedness of *Black Panther* aligns with Okorafor's conceptualization of Africanfuturism. According to Okorafor, Africanfuturism is 'concerned with visions of the future, is interested in technology … is centered on and predominantly written by people of African descent (Black people) and it is rooted first and foremost in Africa'.

Whereas in traditional science fiction, 'the technologically enabled future is by its very nature unmoored from the past and from people of color' (Nelson: 6), *Black Panther* emphasizes the inextricability of the past, the present, and the future, from one another. Much like in the African cosmological apprehension of the fluidity of time, *Black Panther* re-enacts how the present is shaped by the past, and how the future only proceeds from that present which is shaped by the past. In their essay, Ainehi Edoro and Bhakti Shringarpure assert that Wakanda 'is not futuristic enough', stressing that the nation 'is too rooted in an Africa we already know and inhabit, and thus does not manage to take flight into the imaginary' ('Why is the cultural life of *Black Panther* So Derivative?') It would seem, judging by this position, that the future of Africa *must* be something radically unrecognizable, something that bears no shade of resemblance to Africa's present realities. The shortfall of such an imagination of the future, however, is that it is a future that is fractured from *its* past – and such is no future at all. I align with Tolulope Akinwole's reading of futurity in *Black Panther.* As Akinwole has noted, 'the temporal nodes of past, present, and future are dynamic, in a useful way … the past is the future of a period that time has dimmed, just as the present is the future of the past' ('Embodied Masculine Sovereignty, Reimagined Femininity': 148). *Black Panther*'s cartography of temporality, therefore, is consistent with African cosmology, another factor that characterizes Africanfuturism. As Okorafor puts it, Africanfuturism 'acknowledges the seamless blend of true existing African spiritualities and cosmologies with the imaginative'.

Ultimately, *Black Panther*'s politics of futurity serves as an iconography of appreciating the multivalences of global Blackness. The future, as it were, is a Blackness that successfully recoups its 'multiple' consciousness. Since W.E.B. Du Bois promulgated his insightful term of 'double consciousness', the realities of global modernity have become even more deeply complicated and complex. Hence, Blackness – like Whiteness as well as any other term that captures lived racial experiences – cannot escape the complex and multi-layered dimensions of its embodied realities. Importantly, Blackness in diasporic discourses needs to be more nuanced to account for these complexities. What *Black Panther* invites us to acknowledge is that the future of Blackness, especially global diasporic Blackness, largely depends on how Black nations negotiate their sovereignties and how Black people navigate their individual subjectivities – however complex and difficult that may seem like in the present. In that light, the conflict between Killmonger and T'Challa is, really, a contemplation about the prospects and problematics of global Black solidarity, even if it seems on the surface to be a conflict of ideology and principle – of whether Wakanda should be involved in global politics or not.

The clash between T'Challa's nationalism (even if parochial at the beginning) and Killmonger's Pan-Africanist solidarity (even if motivated by personal vendetta) is resolved at the end of the movie in Wakanda's eventual embrace of the world. In his scathing review of *Black Panther*, Christopher Lebron avers that the movie presents its viewers

> two radical imaginings ... an immensely rich and flourishing advanced African nation that is sealed off from white colonialism and supremacy; and a few black Wakandans with a vision of global black solidarity who are determined to use Wakanda's privilege to emancipate all black people. ('*Black Panther* is Not the Movie We Deserve')

Lebron notes, quite profoundly, that 'these imaginings could be made to reconcile'. Even though Lebron's critique is searing and he asserts that 'the movie's director and writer (with Joe Cole), Ryan Coogler, makes viewers choose' between the two imaginings, I opine that what the movie does, rather, is to highlight the perennial tension among Black people all over the world, and the deep complexities of global Black solidarity structures such as Pan-Africanism. Moreover, by charting an existential journey of Wakanda from nationalist chauvinism to a broadminded, Pan-African ethos, *Black Panther* articulates a desire for a certain type of Black internationalism that does not paper over the depth of difficulties that attend cultural movements of global solidarity. The message, in clear terms, is that Africa and its diaspora cannot do without each other if the future of global Blackness is to be productively apprehended.

CONCLUSION

In his edited collection, *Contemporary Speculative Fiction: Critical Insights*, Keith Booker asserts that 'science fiction places readers in a world different from their own (made so by specifically identifiable drivers of change, or "novums", such as technological advances) and then challenges them to examine and critically analyse those differences' (4). Booker notes further that:

> This cognitive process has a potentially powerful political impact in that it encourages readers to reexamine assumptions about their world and to realize that even the most fundamental things about their world might be different. Science fiction thus emerges … as a key utopian form that helps readers to formulate challenges to the status quo and to imagine the possibility of alternatives. (5)

It is this activist work of challenging the status quo that *Black Panther* does.

In this chapter, I have argued that *Black Panther* reconfigures historiographical memories about Africa, by decentring the prominent role that colonialism has taken as a staple topos in prevalent contemplations of African history. Of course, the movie does not seek to write colonialism out of Africa's history – my analysis shows that Wakanda, even though portrayed as an uncolonized nation, actually offers an allegorical testimony to how colonialism upended and continues to upset the internal and external sovereignties of African countries. Instead, *Black Panther* compels the viewer to realize that 'even the most fundamental things about their world', as Booker puts it, such as colonial legacies, imperial neocolonialism, etc., could be viewed through a different critical epistemological lens. This form of 'looking back' to recuperate the past is not merely to conjure up alternative histories; the purpose is to nuance the way we remember, to 're-member' marginalized histories that have been dismembered to suit imperial ends. But most importantly, it is such a nuanced *remembrance* of the past that can engender a profound imagination of the future.

NOTES

1 One of the latest manifestations of this brand of historical revisionism is Bruce Gilley's highly controversial article, 'The Case for Colonialism', which was originally published in *Third World Quarterly* in 2017 but was later taken down following a massive backlash.

2 Isiah Lavender III notes, in *Afrofuturism Rising*, that 'Like science fiction (SF) itself, Afrofuturism proves notoriously difficult to define' (1).

3 Nnedi Okorafor says she coined the term Africanfuturism because 'the term Afrofuturism had several definitions and some of the most prominent ones didn't describe what [she] was doing' ('Africanfuturism Defined').

WORKS CITED

Akinwole, Tolulope. 'Embodied Masculine Sovereignty, Reimagined Femininity: Implications of a Soyinkaesque Reading of Ryan Coogler's *Black Panther*'. *Cambridge Journal of Postcolonial Literary Inquiry* Vol. 7, No. 2, 2020: 147–157.

Asante, Godfried and Gloria N. Pindi. '(Re)imagining African Futures: Wakanda and the Politics of Transnational Blackness'. *Review of Communicatio,* Vol. 20, No. 3, 2020: 220–228.

Booker, Keith (ed.). *Contemporary Speculative Fiction: Critical Insights*. Ipswich: Salem Press, 2013.

Burnett, Joshua Yu. 'The Great Change and the Great Book: Nnedi Okorafor's Postcolonial, Post-Apocalyptic Africa and the Promise of Black Speculative Fiction'. *Research in African Literatures* Vol. 46, No. 4, 2015: 133–150.

Dery, Mark. 'Black to the Future: Interviews with Samuel Delany, Greg Tate, and Tricia Rose', Flame Wars: The Discourse of Cyberculture, special issue of *South Atlantic Quarterly* Vol. 92, No. 4, 1993: 179–222.

Eatough, Matthew. 'African Science Fiction and the Planning Imagination'. *Cambridge Journal of Postcolonial Literary Inquiry* Vol. 4, No. 2, 2017: 237–257.

Edoro, Ainehi and Bhakti Shringarpure. 'Why is the cultural life of *Black Panther* So Derivative?' *Africasacountry*, 26 February 2018. https://africasacountry.com/2018/02/africa-is-a-country-in-wakanda (accessed 15 November 2020).

Fanon, Frantz. *The Wretched of the Earth*. New York: Grove Press, 1961.

Langer, Jessica. *Postcolonialism and Science Fiction*. New York: Palgrave Macmillan, 2011.

Lavender, Isiah, III. *Afrofuturism Rising: The Literary Prehistory of a Movement*. Columbus: Ohio State University Press, 2019.

Lebron, Christopher. '*Black Panther* is Not the Movie We Deserve'. *Boston Review*, 17 February 17. http://bostonreview.net/race/christopher-lebron-black-panther (accessed 17 November 2020).

Mazrui, Ali and Michael Tidy. *Nationalism and the New States in Africa*. London: Heinemann Educational, 1984.

Mbembe, Achille. 'Necropolitics', *Public Culture* Vol. 15, No. 1, 2003: 11–40.

Nelson, Alondra. 'Afrofuturism', *Social Text*, Vol. 20, No. 2, 2002: 1–15.

Okorafor, Nnedi. 'Africanfuturism Defined'. *Nnedi's Wahala Zone Blog*, 19 October 2019. http://nnedi.blogspot.com/2019/10/africanfuturism-defined.html?m=1 (accessed 10 November 2020).

Samatar, Sofia. 'Toward a Planetary History of Afrofuturism'. *Research in African Literatures* Vol. 48, No. 4, 2017: 175–191.

Soyinka, Wole. 'Memory, Truth and Healing'. In *The Politics of Memory*, Ifi Amadiume and Abdullahi An-Na'im (eds). London: Zed Books, 2000: 21–37.

Wabuke, Hope. 'Afrofuturism, Africanfuturism, and the Language of Black Speculative Literature'. *Los Angeles Review of Books*, 27 August 2020. https://lareviewofbooks.org/article/afrofuturism-africanfuturism-and-the-language-of-black-speculative-literature (accessed 10 November 2020).
Womack, Ytasha. *Afrofuturism: The World of Black Sci-Fi and Fantasy Culture*. Chicago, IL: Lawrence Hill Books, 2013.

African Counter-utopias

From Counter-narratives to the Presentification of Alternative Worlds

ÉRIC ESSONO TSIMI

Yambo Ouologuem received the Prix Renaudot in 1968, while he was working as a high school teacher in France.[1] In 1969, crowned with his new prestige, he wrote a 'letter' to the French and Africans in France. Using enthusiastic but often excessive language, the Malian author – who died in 2017 – cast a gloomy, if not pessimistic, and redundant look on Africa. Redundant, in the sense that it took up a certain number of colonial clichés, and that repetition was reflected in its sense of humour, as underscored in the following extract:

> *Le Blanc seul a un nom, le Noir n'a que des surnoms : Nègre, Abibi, Bibine, Boudouboudou, Alibi, etc. C'est pourquoi le tchandathrope, découvert récemment au Tchad, en Afrique, était ainsi nommé parce que, lorsqu'on le déterra, il fit atchoum ! Il avait ainsi craché toute la noirceur de son âme, ce qui permit de conclure que c'était, non point un Blanc, mais bien un vieux Nègre.* (Ouologuem, *Lettre à la France nègre*: 46)
>
> [Whites alone have a name, Blacks only have nicknames: Negro, Abibi, Bibine, Boudouboudou, Alibi, etc. Therefore, the Tchandathrope,[2] recently discovered in Chad, Africa, was so named because, when unearthed, he sounded like a shock! He had thus spat out all the darkness of his soul, which led to the conclusion that he was not a White, but an old Negro.][3]

In his *A Black Ghostwriter's Letter to France*, the author of the controversial *Le devoir de violence* ('The Duty of Violence') thought he was being humorous to the detriment of Africans (or some of them), notably Léopold Sédar Senghor, who was full of praise for the German ethnologist Léo Frobenius (1873–1938). According to Senghor ('Les leçons de Léo Frobenius') it was Frobenius, more than any other, who illuminated for us, words like 'emotion', 'art', 'myth', 'Eurafrica'. In the fourth and fifth chapters of the section entitled 'La nuit des géants', we meet the character of Fritz Shrobenius, a learned buffoon who resembles Léo Frobenius.

Ouologuem criticizes the work of Frobenius and the essentialism of the writers of negritude, thus mocking Senghor at length. Likewise, in Ouologuem's 'letter', which was advertised as a pamphlet, he repeats outdated perceptions and clichés, which appear humiliating and racist. The English translation, *A Black Ghostwriter's Letter to France* (Wise, 2008) does not use the word 'nègre' and instead uses one of its many meanings corresponding to 'ghostwriter'. Another discrepancy which departs from the French title is in the use of the singular in 'letter', emphasizing the epistolary genre, when in all, there are about thirteen letters, or more precisely thirteen satirical essays by Yambo Ouologuem on the 'black problem'. The chapter titles include (I translate): 'Letter to mixed couples', 'Letter to black kings passing through France', 'Letter to negatively [nègrement] single women'. This last is a fierce assault on Whites and Blacks, Christians, Muslims, and Animists; everyone becomes the object of his frustration, insofar as he dismisses both Whites and Blacks. The initial reference to Yambo Ouologuem is not trivial. *Le devoir de violence* is a fresco that extends from the thirteenth to the twentieth centuries. It tells the fate of Nakem and the Saif dynasty's imaginary empire, who reign there as devious masters. Yambo Ouologuem shows the complexities of Africa's history where slavery and colonization would predate the arrival of Europeans who would only have taken over and dramatically amplified an already existing crazy system.

The imaginary West African empire of Nakem that Ouologuem creates is a reference to Kanem-Bornu. Nakem becomes Kanem in verlan.[4] The allegorical kingdom described in *Le devoir de violence* is typical of the treatment of imaginary places in African counter-utopias (Waberi, *Aux États-Unis d'Afrique*; Amoussou Sauvil, *Africa paradis*; Miano, *Rouge impératrice*). This genre consists of inversions; a questioning of the place of the non-place as, etymologically speaking, 'utopia' signifies 'no-place'. *Le devoir de violence*, condemned by Léopold Sédar Senghor in his time, can be seen not only as a counter-utopia, but also a 'memorial utopia' (Coquio, *Le mal de vérité*). The memorial utopia presents the idea that the past, already ruined by the absence of a future, resuscitates (from) the ruin of the future. Now presentified, the experience of the past is itself part of futurition; in our effort to bring about 'anew' the appearance of an old experience results in a new venture that can be said to be counterfactual or, where appropriate, counter-utopian. The advocates of negritude resented the virulent denunciation of the role of Africans in the slave trade: Senghor reproached Ouologuem for 'denying his ancestors', however at publication, the critical reception in France was mitigated (Orban, *Livre culte, livre maudit*).

African writers living in France are subject to the same external influences: the climate, prejudices, and organic racism of society, and so they react to the same cultural stimuli. For example, these lyrics by Claude Nougaro covering Louis Armstrong (*Go Down Moses*), demonstrate the interplay of common influences on authors who all drank from the sources of Lethe.[5] The following song, translated from French, can be found in Abdourahman Waberi (*Aux États-Unis d'Afrique*: 69):

> Armstrong, I'm not black
> I am white-skinned
> When we want to sing about hope
> What unfortunate bad luck.
> Yes, I can see the sky, the bird,
> Nothing, nothing, nothing shines up there
> The angels ... Zero
> I am white-skinned

This song is later cited by Mabanckou in *Black Bazar* and by Miano in *Blues pour Elise* (27)

Africa's 'false start' (Dumont 'False Start in Africa') is presented in 'Nakem' by Yambo Ouologuem in 1969. Judging by the lexicographic occurrences of 'darkness' in many Francophone African authors' work as well as in the work of Mabanckou, this Africa seems doomed to an endless curse. How then, is such an Africa said to be damned, locked up in 'the dark night' (Mbembe, *Out of the Dark Night*)? Does the imagination of the arts and letters create places of dreams or the impossible? How do you see it and defend the best of all possible worlds? *Aux États-Unis d'Afrique* (In the United States of Africa) recounts a Wakanda-like Africa: The Federation of the United States of Africa thrives with its wealthy centres. Still, it remains indifferent to the plight of millions of people from Europe and America, desperately trying to take refuge on its lands. We can then cite *Aux États-Unis d'Afrique* and the film *Africa Paradis*, to show that the contemporary African counter-utopia is a presentification. That is to say, it is a fusion of the imaginary and the counterfactual, in determined referential spaces; places loaded with meaning. As in a Voltairian journey, the Francophone African authors seem to be like Pangloss, whose best of all possible worlds, the El Dorados, are always reactivations, 'presentifications', of the surrounding chaos: inverted mirrors of order. What takes the place of a dystopia is a reinvention of Africa from the ashes. All the obscure references which in the mind of the public refer to Africa, the Mater Dolorosa of humanity, are then redacted. In *Aux États-Unis d'Afrique* and *Africa Paradis*, the counter-utopia constitutes then a signifying system, and an intermedial discourse (cinema, literature, and philosophy), which unfolds under the double relation of discursive strategies (how the artist will articulate) and cognitive processes (what s/he imagines, thinks, and feels) aimed at anticipating the decline of the West. *Aux Etats-Unis d'Afrique* is documented, researched fiction, which draws its narrative identity from the names of real places and people, from facts and contemporary creations reminiscent of Africa and its fragmented history (continental Africa and Africa of the diaspora).

On the back cover, the novel is described as an 'epic' and a 'pamphlet'. Like *Le devoir de violence*, this epic can be called a counter-epic, this utopia a counter-utopia: Blacks are slavers like any other and Kanem is Nakem. We are in a paradigm shift which employs the logic of reversed perspectives. This 'transgeneric' work is fragmented into four very unequal parts. The fourth part, *The Return to Asmara*, has only one chapter, which closes the thirty-one chapters that opened with a *Journey to Asmara*. The '*forme-sens*' of critical circularity corresponds to a movement around oneself rather than a revolution around Asmara. The articulation between form (the reading protocol) and meaning (the narrative programme) and the coupling of these two concepts that we owe to Saussure and Meschonic, allows us to consider Waberi's text as a regulated space of scattering of counter-utopian ideas. The self remains in the background; it is never 'I'-positioned. The novel is narrated in the second person (you) and even at times in an impersonal way (we, he, one). From the first page of the text, the stakes are raised, and the racialized dynamics are properly announced. The 'title' of the first chapter is:

> *Où l'auteur rend un compte succinct des origines de notre prospérité et des motifs qui ont jeté les Caucasiens sur le chemin de l'exil.* (*Aux États-Unis d'Afrique*: 11)
>
> [Where the author succinctly recounts the origins of our own prosperity and the reasons why the Caucasians have been thrown onto the path of exile. (*In the United States of Africa*)]

The narrator often distances themself from the author. The author, for his part, often identifies with the reader (cf. underlined words) and writes in a permanent blurring of narrative instances and a travesty of auctor / lector functions:

> *Ce Caucasien d'ethnie suisse parle un patois allemand et prétend qu'il a fui la violence et la famine de l'ère du jet et du net … Appelons-le Yacouba, primo pour préserver son identité, deusio parce qu'il a un patronyme à coucher dehors.* (11)
>
> [This Caucasian of Swiss tribe speaks a dialect of German and claims he's been fleeing from famine since the time of Net … **Let's call him** Yacouba, first to preserve his identity and, second, because his last name is unpronounceable.]

On the back cover of the book, the publisher JC Lattès speaks of a 'mirror extended to the West', nevertheless the fact remains that this Stendhalian mirror is concave or convex in nature, depending upon the case: it reflects, deforms, enlarges, or the opposite may be the case. In this aesthetic or disaesthetic approach of overturning codes and utopia, the characters' names emerge as a place of identity; Africa, as an unexpected utopia (33–35). The portrait of Malaika (Maya) is drawn with composite, heterogeneous strokes, through which her hybridity is told to us. Every line is a description in this novel, up to and including the narration. Everything is very quickly described with a mass of references which densifies the text but not the plot. The narrator is in the shoes of a storyteller. The narration device which uses the personal pronoun 'you' to designate the main character, Maya, transparent and silent, borrows greatly from orality. To understand the narrator's audience(s), to know 'for whom' he writes, perhaps we first need to take a detour and answer the question of where the author writes? If Waberi is not writing an ex-cathedra novel, he constructs a demanding, even elitist work, confident in the intelligence of the universal narratee, with whom he is familiar, by creating familiarity or a family feeling. From this point of view, this utopia-mirror stands out as a counter-utopia-reflection since, published in France, it addresses the what-if scenario of the West's decline. As a privileged mode of critical representations of the French-speaking world, torn apart by terrorist attacks and recomposed by a large Muslim community, the counter-utopia is undoubtedly the place of choice for the renewal of the novel. The counter-utopia is an objective variant of utopia: the difference lies in its worthiness being that it is not the absolute best but the relative best that one imagines. Diegetic plausibility and the constitution of enunciative devices provide the main axes of interpretation of counter-utopian fictions.

The subject of *Aux États-Unis d'Afrique* is both political and cultural; the themes are identity and migration. Having said that, we can see that this chronicle is uchronic in its representation of the 'fantasy' that

Black people would have taken, if not the skin colour, then at least the place of the White people. It depicts a world turned upside down: one that is counter-utopian in nature. This version of reality is also reminiscent of a mirror, presenting a specular image of reality. Its merit consists of bringing to life an African heritage in all the diversity of its learned and artistic production. *Aux États-Unis d'Afrique* is fiction without heroes. The author does not want to share the psychological evolution of Maya that would normally be expected over the course of a narrative/plot. The paradigm of the quest for identity is prominent, but the crumbling of the narrative does not make it possible to work in an in-depth manner on this character who is spoken of more than she herself speaks. The reader does not see Maya's passage through history. Instead, the reader goes through history in Maya's shoes, thanks to the author's enunciative, polyphonic device. It is also counter-utopia in the sense that the story (its place) is supposed; it is fantasized about rather than said, akin to a lengthy interior monologue, in a text from which the plot is almost absent. However, the epigraphs at the head of the chapters are reminiscent of adventure stories and are precious reference points for understanding. Maya is a pretext, used as a generic, translocal figure with a fluid spatial identity. Maybe Maya is just a character for meeting the novel's requirement of having a lead character. Everything here is masks and shadow play, including, first of all, the place of 'utopia'.[6]

The author brings about a reversal of perspectives. Asmara becomes the centre of the world, with Africa as the centre of all significant economic, intellectual, and military operations. McDonald's is called McDiop, Pepsi becomes Papsy, but unlike a utopia, places exist even though the geopolitical configuration and contemporary setting make the sites unreal, and the situations described historically improbable. Moreover, the labelling of the book in the 'novel' genre remains questionable as the text does not comfortably fit within that category. The author of *The Right Stuff*, Tom Wolfe, has termed this halfway state between fiction and non-fiction as 'New Journalism'. In *Aux États-Unis d'Afrique*, Yacouba is a Helvetian whose name was Maximilien Geoffroy de Saint-Hilaire. He fled his country (Switzerland), where they kill each other whenever and wherever they can. Yacouba finally meets a sordid death in Africa. Besides him, there is a silent character (who is never heard) to animate this fiction: Maya, a young girl who had the misfortune to come into the world in a *banlieue d'un bidonville* in Normandy (Waberi: 22). 'Yacouba the carpenter' is reminiscent of Jacob (a.k.a. Israel) or Job; these biblical characters are figures of exile.

There is a blurring of boundaries between fantasy, documentary, and historical archives. It is a unique book that is comparable to a tale, a fable or even literary farce. Maya is an adopted 'Caucasian'; Yacouba is a common first name, used on occasion to avoid pronouncing *un nom à coucher dehors* (an unpronounceable name) of a Caucasian, Western migrant. The author's writing is poetic, precise, and often jubilant; it corresponds to the substance he wants to convey; it is part of his 'project'. There are no big surprises apart from the aesthetic pleasure of seeing the well-chosen words line up.

The idea of this 'outside world, this Africa so hardly African, and this black phoenix, which is reborn from white ashes', does not originate with Waberi. No, such a utopia was not new and indeed already existed, even theoretically. The enunciative masks that the author uses by making the narratee the other hero of his novel: 'You' in place of the 'I', of 'he', or of the impersonal 'we' are not new either and, in point of fact, are typical of African counter-utopian aesthetics (Ehora 'Les "nouveaux habits" de l'oralité chez les romanciers ouest-africains de la seconde génération': 30). As if by not naming Africa the place of utopia, by anonymizing it, the West would be deemed the site referred to, as popular imagery tends to do. The film *Africa Paradis*, released in 2006 – the same year as *Aux Etats-Unis d'Afrique*, has a synopsis which reads as follows: 'In an imaginary future, Africa has entered an era of great prosperity, while Europe has sunk into poverty and underdevelopment' (np). It is also about the United States of Africa, a prosperous country. The story recounted is that of a couple from France, a country which has become a shadow of itself. From the start of the film, we discover an apocalyptic Paris, so close to the suburbs of Les Misérables. The gates of paradise are located at the Embassy of the United States of Africa, where the French visa applicants go at dawn in the hope of obtaining the precious key. The comical and subversive scene shows Olivier, an engineer, and Pauline, a teacher, as candidates for exile. Immigration officials offer them jobs as labourers or housekeepers. He tells them that the United States of Africa already has prestigious schools training their own engineers and teachers. Simultaneously, in the United States of Africa, members of parliament (MPs) are rallying around the woman president to discuss new immigration laws. In a spaceship-like parliament, two MPs clash: Yokossi, from the African Radical Party, an 'anti-white extremist', and Koudoussou, from the African Liberal Party, which wishes to integrate Europeans into African civic life. In Paris, the couple, deprived of visas, resort to a smuggler's services and land illegally in the United

States of Africa. They are then transferred to a transit camp pending their deportation. Olivier runs away from the camp and Pauline becomes the servant of Deputy Koudoussou. In the film, we spin the metaphorical poles more elaborately. We remember that Waberi went on to develop a concept of Africa existing as a federation with the author's homeland as the world's centre.

The African counter-utopia mocks utopia: counter-utopia this side of the Pyrenees, utopia beyond.[7] In *Aux États-Unis d'Afrique*, the mark of an African author can be found. We see Waberi committed to the causes of his country of origin while, at the same time, he defines himself as 'a French writer'. He claims he belongs to a universal 'interpretive community' and, consequently, finds nuances of belonging inessential (African, Djiboutian, or French, for example). Perhaps the categories utopia/dystopia/counter-utopia offer reading grids to the critic, but they do not bear any consequence to the reader. It is from a place of epistemic commitment that a writer will treat a particular subject in a particular way. Abdourahman Waberi makes counter-utopia a genre in motion, notably through the story of a quest and an investigation into identity which reveals great intertextual richness. Prosperity has changed sides, and El Dorado is no longer in the West; instead, it is in Africa. Presenting counter-utopian aesthetics does not consist of bringing the past to life, nor does it consist of anticipating the near future (in the sense of updating both the past and the future). It consists of bringing the non-place to life. Therefore, the counter-utopia is an actualization of the diseases of idealities that project themselves elsewhere: El Dorado, a utopia in the West, is as inaccessible as it is imaginary in most Africans' eyes. By imagining a world in which young Europeans are desperately trying to reach Africa, is Beninese filmmaker Sylvestre Amoussou building an ideal world for anyone? In *Africa Paradis*, we see a French couple attempting, in 2033, to flee a Europe plagued by unemployment and violence, to enter into the prosperity of the United States of Africa. The African border guards stop them, and it is the miseries that African migrants are currently experiencing that are found projected there. While the film has been described as a utopia (Lawrence, 'Difficult Satire under Austerity': 131), it is in fact a counter-utopia just like Waberi's novel.

Going to Africa is within reach of many European or North American citizens. It is not as inaccessible as Europe or America are for Africans. My American students find it normal to spend their summer holidays in Morocco, Egypt, or South Africa in order to discover both the world and themselves. Still, they often do not realize the difficulties experienced by African nationals who are mobile. The counter-utopias that show, imagine, write, and think about Africa as a migratory paradise not only help correct stereotypes about Africa but challenge the very meaning of utopia. *What if?* has become a staple question in new African fiction. The African counter-utopia needs to be distinguished from both dystopia and Afrofuturism, crossed as it is by an aporetic aesthetic of the presentification of a non-place of the ideal. The presentification of historiographical absence and 'no-place' consists of an aesthetic of narrative eccentricity. The counter-utopia is then a presentification, which leads to a new experience of historicity and historical facticity, and is subsequently redefined as fiction. Waberi's, Miano's, and Amoussou's counter-utopias reinvent the scope of utopia by fusing past and future times. Dreamed and distant places are repatriated into improbable tales, ultimately justifying the counter-utopian meaning and value of Africa, if not as a 'place', then in any case as an unlikely 'object'. This trend is confirmed by the Ghanaian film, directed by Kobi Rana, which while entitled *Utopia* (2016), only relates to the drifting youth who in their immense efforts to achieve their goals, are prepared to make the worst compromises to achieve happiness. Theirs is a eutopia; a counter-utopia rooted in the real world.[8]

Summing up this chapter in the form of two central points that are both open to discussion, I would like to maintain that, first, the counter-utopia is a new art of speaking about a present that escapes us through a possible future. Francophone counter-utopian works are distinct from Anglophone dystopias. The counter-utopias talk of a future time that is lost in the present times. They pose literature as an epistemological tool that either makes many aspects of society intelligible or questions its unintelligibility. On the difference between counter-utopia and dystopia: there are noticeable slippages in meaning when counter-utopia is designated as a simple synonymic variant of dystopia, positioning it as a paragon of linguistic prosthesis. The Italian philosopher, Gianni Vattimo, author of the article 'Toward An Ontology of Decline', estimated in a recent article that 'one of the essential characteristics of utopia today is the rise and dissemination of a literary genre that has been variously defined as anti-utopia, dystopia or counter-utopia'. Patrick Hubner (2015: 1157) noted: 'Just as the myth engenders its counter-myth through an inversion of the values of the original structure – the myth of the ideal came to dominate, under the various appellations of "counter-utopia," "anti-utopia," "dystopia" or indeed "cacotopia"!' Counter-utopia is a concept in a state of relative indefiniteness. If utopia is considered a literary genre, its 'inverted double' (Godin 'Sens de la contre-utopie'), the counter-utopia seems, of recent appearance, an ultra-contemporary fashion. Mark Lilla ('Slouching Toward Mecca') had foreseen from the publication of *Soumission* that Michel Houellebecq had created a new 'genre',[9] a genre anchored in the idea that we live 'in the ruins of the future' (DeLillo, title). This genre, which he had not named, is what the Francophone African presentification of alternative worlds is all about.

Second, the examination of the African counter-utopia is, therefore, an opportunity for literary research to develop its own discourse on subjects shared with other academic disciplines: identity, progress, time, history, and the future. Within the humanities, research into the counter-utopia will continue to expand, especially in these troubled times marked by a more fluid circulation of ideas between literary texts, texts from popular culture, and more experimental fiction.[10]

NOTES

1 Prix Renaudot is a French literary prize awarded to the author of an outstanding original novel. It is one of the most prescriptive accolades for book sales. Five sub-Saharan novelists have received it to date, namely: Yambo Ouologuem (1968), Ahmadou Kourouma (2000), Alain Mabanckou (2006), Tierno Monénembo (2008), Scholastique Mukasonga (2012).

2 In 1965, Lucy's co-discoverer, anthropologist Yves Coppens, found at the foot of the Angamma cliff in Chad a heavily eroded skull, which he ventured to call Tchadanthropus. It would be a *Homo erectus*. But, here, scientific precision matters less than the ironic aim of Ouologuem, which thus echoed in his book a scientific topicality.

3 All translations are by the author of this chapter, unless otherwise noted.

4 Verlan is a type of argot in the French language, featuring inversion of syllables in a word, common in slang and youth language.

5 Claude Nougaro (1924–2004) is a French singer-songwriter and poet. The rather grim view he had of the world coexisted with an overwhelming urge to hope. He admired the United States, loved jazz and Black artists. For this reason, perhaps, some authors from French-speaking Africa have nurtured sympathy for him, to the point of interpreting some of his lyrics as a racial counter-utopia, so to speak.

6 While 'counter-utopia' is the genre, the place can be referred to as a utopia, which then acts as an umbrella term.

7 I distinguish classic dystopia from counter-utopia. Utopia has been inhabited in a new way (as the Latin saying goes, '*non nova, sed nove*' – Not new things, but in a new way) while it is referring to a place which does not have any

location, being only a creation of the author's imagination, whereas counter-utopia is referring to a 'topos', an actual place, Africa in this case.

8 Thomas More used the term 'eutopia' to refer to the imaginary place he had designed. This neologism is no longer based on the negation or but on the Greek prefix EU, which one finds in euphoria and which means 'good'. Only the eponymous term 'utopia' has passed down to posterity.

9 'Michel Houellebecq has created a new genre – the dystopian conversion tale. *Soumission* ('Submission') is not the story some expected of a coup d'état, and no one in it expresses hatred or even contempt of Muslims. It is about a man and a country who through indifference and exhaustion find themselves slouching toward Mecca. There is not even drama here – no clash of spiritual armies, no martyrdom, no final conflagration. Stuff just happens, as in all Houellebecq's fiction. All one hears at the end is a bone-chilling sigh of collective relief. The old has passed away; behold, the new has come. Whatever.' Mark Lilla ('Slouching Toward Mecca'. This passage appears as such in his book (Lilla, *The Shipwrecked Mind*: 117)

10 I am grateful to the anonymous reader who suggested numerous edits and precisions throughout the paper. I also thank David Lynch, who proof-read the draft before its submission. Thanks to this issue's co-editor, Louisa Egbunike, too.

WORKS CITED

'Africa Paradis'. *ALLOCINÉ*, 2007. www.allocine.fr/film/fichefilm_gen_cfilm=124762.html (accessed 15 April 2021).

Amoussou, Sylvestre and Pierre Sauvil. *Africa paradis*. Rosny-sous-Boi: Tchoko Tchoko 7ème art, 2006.

Coquio, Catherine. *Le mal de vérité: ou l'utopie de la mémoire*. Paris: Armand Colin, 2015.

DeLillo, Don. 'In the Ruins of the Future'. *The Guardian*, 22 December 2001.

Dumont, René. 'False Start in Africa', trans. Phyllis Nauts Ott. New York: Praeger, 1969 [1966].

Ehora, Effoh Clément. 'Les "nouveaux habits" de l'oralité chez les romanciers ouest-africains de la seconde génération'. In *Littérature africaine et oralité*, Ursula Baumgardt (éd.). Paris: Karthala, 2013: 29–52.

Godin, Christian. 'Sens de la contre-utopie'. *Cités* Vol. 2, No. 42, 2010: 61–68.

Houellebecq, Michel. *Soumission*. Paris: Flammarion, 2015.

Hubner, Patrick. 'Utopia and Myth'. In *Companion to Literary Myths, Heroes and Archetypes*, Pierre Brunel (ed.). London: Routledge, 2015.

Lawrence, Andrew. 'Difficult Satire under Austerity: Three Films by Sembène, Sissako and Amoussou'. *L'Archéologie du Futur: Cinémas Africains et Utopies* (Archaeology of the Future: African Cinemas and Utopias), Viviane Azarian, Ute Fendler, and Aminata C. Mbaye (eds). München: Akademische Verlagsgemeinschaft, 2018: 139–151

Lilla, Mark. 'Slouching Toward Mecca', 2 April 2015. www.nybooks.com/articles/2015/04/02/slouching-toward-mecca (accessed 11 June 2021).
—— *The Shipwrecked Mind: On Political Reaction*. New York Review of Books, 2016.
Mabanckou, Alain. *Black Bazar*. Paris: Éditions du Seuil, 2009; English translation – London: Serpent's Tail, 2010.
Mbembe, Achille. *Out of the Dark Night: Essays on Decolonization*. New York: Columbia University Press, 2021.
Miano, Léonora. *Blues pour Élise*. Paris: Plon, 2010.
—— *Rouge impératrice*. Paris: Grasset, 2019.
Nougaro, Claude. 'Armstrong'. *Bidonville*. Phillips. 1963
Orban, Jean-Pierre. 'Livre culte, livre maudit: Histoire du *Devoir de violence* de Yambo Ouologuem'. *Continents manuscrits*, 2018. http://journals.openedition.org/coma/1189 (accessed 11 June 2021).
Ouologuem, Yambo. *Le devoir de violence*. Paris: Seuil, 1968
—— *Lettre à la France nègre*. Paris: E. Nalis, 1969.
—— *Bound to Violence* [*Le devoir de violence*], trans. Ralph Manheim. New York: Harcourt Brace Jovanovich, 1971.
—— *The Duty of Violence, A Black Ghostwriter's Letter to France, and The Thousand and One Bibles of Sex*, trans. Christopher Wise. Trenton, NJ: Africa World Press, 2008.
Senghor, Léopold Sédar. 'Les leçons de Léo Frobenius' *Présence Africaine* No. 111, 1979: 142–151.
Vattimo, Gianni. 'Toward an Ontology of Decline'. In *Recoding Metaphysics*, Giovanna Borradoria (ed.). Evanston, IL: Northwestern University Press, 1988: 63–75.
Waberi, Abdourahman A. *Aux États-Unis d'Afrique*. Paris: JC. Lattès, 2007.
Wolfe, Tom. *The Right Stuff*. New York: Farrar, Straus & Giroux, 1979.

Shifting the Frame

Re-imagining Chinua Achebe's *Things Fall Apart* & *Arrow of God* as Speculative Narratives

CLARA IJEOMA OSUJI

INTRODUCTION

There has been a longstanding tradition of inventing 'other' worlds in literary writings. In Chinua Achebe's *Things Fall Apart* and *Arrow of God*, there are representations of supernatural events that have a grounding in the cosmology of the Igbo people. Thus, the author appears to engage in the speculative, based on the rich folklore of south-eastern Nigeria. However, can the creative specialization(s) of Achebe in these two narratives be convincingly factored into the equation of speculative fiction? What really qualifies as speculative fiction? The term speculative fiction has been explained by different scholars with diverse, sometimes, parallel and overlapping views; but, in most of the definitions, there appears to be an attempt to goad speculative fiction (spec-fi) and its related genre, science fiction (sci-fi), on a race for supremacy. Writers such as Margaret Atwood differentiate speculative fiction from science fiction by explaining that speculative fiction refers to a story that 'could happen but hasn't yet [happened]' while science fiction dwells on stories of 'fantasy [that] could not happen' (*In Other Worlds*: 7). Then, can the supernatural events in *Things Fall Apart* and *Arrow of God* happen in real life?

While Atwood explains speculative fiction as stories that are capable of happening in real life, Diana Waggoner defines it as a broad category of modern literature that treats 'supernatural and/or nonexistent phenomena (such as the future) as a special class of objectively real things or events' (*The Hills of Faraway*: 9). Also, Judith Merril states that speculative fiction 'makes use of fantastic and inventive elements to comment on, or speculate about society, humanity, life, the cosmos, reality [a]nd any other topic under the general heading of philosophy' ('Introduction': 3). While Atwood offers a narrower idea of speculative fiction, Waggoner and Merril's definitions are more encompassing.

However, the Dutch bilingual novelist and poet, Sébastien Doubinsky, admits that capturing the term, speculative fiction within a definition is problematic. Nonetheless, he offers an interesting explanation:

> Speculative fiction is a paradox. Synonymous with science fiction and 'genre literature,' it is also one of the most ancient modes of storytelling in literary history … There is a distinct speculative thread throughout literary narrative that can be traced from the origins of writing until the present day. ('Jordan Krall's Speculative Fiction': np)

Doubinsky makes an interesting reference to a manifest 'speculative thread' that commonly exists in literary fiction. Thus, in identifying particular episodes in Achebe's *Things Fall Apart* and *Arrow of God* for close reading, the 'speculative threads' in the two novels are examined. Remarkably, it is the vague aspect of speculative fiction which Doubinsky also highlights in the passage that makes the genre malleable and more open to different explanations and applications. Hence, for Michael Svec and Mike Winiski, spec-fi simply describes a more contemporary application of the term to prose fiction which offers a 'fascinating mix of science, personal experience and fantasy' ('SF and Speculative Novel': 43). The reference to personal experience and fantasy resonates with Achebe's Igbo socio-cultural beliefs and worldview which he keenly engages and represents as realities not only in these two novels but in works such as *Anthills of the Savannah*.

Although Doubinsky's idea of the 'speculative thread', to a large extent guides the analysis of *Things Fall Apart* and *Arrow of God* in this chapter, nonetheless, Svec and Winiski's outlook on the genre is also useful. This is because examining the speculative in Achebe's texts from Svec and Winiski's perspective, challenges the reader to ask what *Things Fall Apart* and *Arrow of God* could divulge about the speculative fictional artistry of Achebe. Perceptibly, speculative fiction is a wide-ranging term which focuses on prose narratives with supernatural, folkloric, magical realist, futuristic, and fantastical phenomena. Thus, looking for a considerably deep content of the speculative in Achebe's selected novels, tests the reader's conceptual understanding and ability to discern speculative elements in otherwise 'serious' novels. Undoubtedly, in these novels, Achebe invokes remarkable realistic communities where deities, ancestors and other spiritual personages interact with human characters. This is a world that is, largely, shaped by spirit beings who traffic seamlessly through time and space, exacting justice and directing the affairs of Umuofia and Umuaro.

THINGS FALL APART AND *ARROW OF GOD* THROUGH THE LENS OF SPECULATIVE FICTION

While Achebe's classic, *Things Fall Apart*, and his third novel, *Arrow of God*, have been subjected to various intellectually convincing interpretations, this chapter attempts to further subject the texts to a new demand by linking them to African speculative fiction. The speculative elements in the two texts, mostly revolve around the characters of Chielo (the priestess of Agbala) and the *egwugwu* (the nine spiritual fathers of the clan) in Achebe's fictional community of Umuofia in *Things Fall Apart*, and the deity Ulu and his chief priest, Ezeulu, in *Arrow of God*. The question is, by placing these novels within the frame of speculative fiction, can they truly speak to the reader through a speculative lens? Paul Kincaid maintains that '[e] verything is capable of being read in different ways. So, the way we read a work, sometimes the way we choose to read a work, is crucial in determining how we identify it' ('Against a Definition of Science Fiction': np). Thus, evaluating the two narratives and attempting to identify speculative elements in them may seem like putting the narratives under pressure in order to fit them within the speculative frame. However, a closer examination reveals that the novels appear to open up more speculative threads than may have been intentionally embedded in them.

Things Fall Apart has proved to be a piece of fiction that speaks to the literary analyst in various genres of fiction and appears to be a salad bowl of ideas where discerning interpretative scholars and literary critics dip their hands to choose and justify their own ideas. Although, the text has often been read as a cultural narrative which details the way of life in Umuofia before the colonial encounter, there is an attempt at a re-reading of some of the events in the novel which read like a cross between spec-fi and sci-fi. For instance, the priestess Chielo is introduced in the sixth chapter of the novel during a wrestling match in Umuofia. This is when Ekwefi, one of Okonkwo's wives, starts a conversation with her as the drummers stop for a brief rest before the real wrestling matches resume. Chielo is a widow, a single mother and a petty trader in her community. Although she could easily be waved off as an ordinary Umuofia woman, the reader gets to know that she is the priestess of Agbala, the famed Oracle of the Hills and the Caves. Consequently, Chielo's life vacillates between the normal (ordinary) and the paranormal (spiritual). As one of the spectators at the wrestling contests, it is evident that aside her priesthood status and roles, she lives a normal life. The reader is informed about the nature of this woman who embodies the spirit of the Oracle in Umuofia:

> In ordinary life Chielo was a widow with two children. She was very friendly with Ekwefi and they shared a common shed in the market. She was particularly fond of Ekwefi's only daughter, Ezinma, whom she called 'my daughter'. Quite often she bought bean cakes and gave Ekwefi some to take home to Ezinma. Anyone seeing Chielo in ordinary life would hardly believe she was the same person who prophesied when the spirit of *Agbala* was upon her. (36)

Unlike her male counterpart in *Arrow of God*, Ezeulu, said to be half-human, half-spirit, Chielo does not have such appellation attached to her status as a priestess. She is simply presented as a widow who acts as the priestess of the Oracle of the Hills and the Caves. However, the significance of the portfolio of the priestess of Agbala can be readily estimated from the role(s) of the Oracle she serves – a highly dignified deity who must examine the legitimacy of all wars before the community can engage in them. This role is even better appreciated in the light of its importance in a war-like community such as Umuofia where the power of this Oracle is palpable. For example, the Oracle dictates that the clan must always attempt a peaceful resolution before resorting to any form of war. The community members understand the fact that they are doomed if their deity refuses to support them if they contravene her injunction and engage in unjust wars:

> In fairness to Umuofia it should be recorded that it never went to wars unless its case was clear and just and acceptable as such by its Oracle – the Oracle of the Hills and the Caves. And there were indeed occasions when the Oracle had forbidden Umuofia from waging a war. If the clan had disobeyed the Oracle they would surely have been beaten, because their dreaded *agadi-nwayi* [old woman] would never fight a war of what the Ibo call *a fight of blame*. (10)

The personality trait of Agbala is highlighted in human terms by endowing her with resoluteness in meting out punishment to those who break the rules of warfare in Umuofia. From the excerpt above, the Oracle comes across as a just and fair deity. In chapter eleven, Chielo's supernatural abilities and spirituality become clearer on the night she visits Okonkwo's compound and demands to leave with his favourite child, Ezinma, at the instance of Agbala. The author seems to consciously heighten the speculative element by preparing the reader for Chielo's arrival to Okonkwo's compound at a time when '[t]he night was impenetrably dark … as black as charcoal' (70). The clear sound of Chielo's voice, her boldness and ability to move swiftly in the 'impenetrably dark' night are indications of her supernatural capabilities as a priestess. From a considerable distance, even before Chielo's arrival to Okonkwo's compound, her prophecy is heard in 'loud and high-pitched voice [which] broke the outer silence of the night' (73). She is biologically female, therefore, her lone thunderous voice while on spiritual duties suggests a further supernatural dimension to her personality that is reminiscent of the 'half-man, half-spirit' attribute of Ezeulu in *Arrow of God*. For indeed, she comes as the mouthpiece of the deity and promptly declares her assignment to Okonkwo. Although, Chielo relates cordially with Ekwefi and is also very of fond of her only child, Ezinma, this night she clearly separates friendship and her communal obligation. This is because when the priestess performs her spiritual roles, her prime focus is on the deity she serves and she is compelled by the exigency of her office to assumes a spiritual dimension as a half-human, half-spirit being.

The reader absorbs the atmosphere of this 'impenetrably dark' night through the reaction of Ekwefi who is now beginning to wonder if she had taken the right decision in following the priestess to the abode of the Oracle of the Hill and Caves. Through Ekwefi's eyes, the reader sees that:

> The world was now peopled with vague, fantastic figures that dissolve under her steady gaze and then formed again in new shapes. At one stage Ekwefi was so afraid that she nearly called out to Chielo for companionship and human sympathy. What she had seen was the shape of a man climbing a palm tree, his head pointing to the earth and his legs skywards. But at that very moment Chielo's voice rose again in her possessed chanting, and Ekwefi recoiled, because there was no humanity there. It was not the same Chielo who sat with her in the market and sometimes bought bean-cakes for Ezinma, whom she called her daughter. It was a different woman – the priestess of Agbala, the Oracle of the Hills and Caves. (78)

In the speculative episode above, there are depictions of a number of seeming alternate realities. For instance, sights that have been familiar to Ekwefi now seem unfamiliar in the night. She sees 'vague fantastic figures' that transform and dissolve as she watches them. One of these figures appears to be a human spirit climbing a palm tree with his head pointing to the ground while his legs are skywards. The silence in the night, apart from the 'ringing' of Chielo's now 'inhuman' distant voice through the neighbouring villages and the myriad grotesque shapes that form and disappear, terrify Ekwefi. While these seem like magical or surreal circumstances that function within the real world of Umuofia, Achebe had signalled that '[d]arkness held a vague terror for these people, even the bravest among them. Children were warned not to whistle at night for fear of evil spirits' (8). Thus, through this speculative thread, Achebe engages with the worldview of Umuofia people and presents this world through the discernments of Ekwefi. The author's timeless approach to crafting stories enables him to place scenes that operate under magical rules to appear as realistic communal experiences.

Chielo is an enigmatic character. As a priestess, she has self-gender-altering abilities, thus, she 'was not a woman that night' (when she carried Ezinma) (79). She has two lives – an 'ordinary life' as a widow, a single mother and a petty trader, and an 'extraordinary life' as the priestess of the legendary Oracle of the Hills and the Caves. As a result, Chielo's life wavers between the normal, a simple widow and one of the spectators of the wrestling contests (35), and the paranormal, a mysterious priestess, whose lone fierce 'loud high-pitched voice' breaks 'the outer silence of the night' (73) as she performs her priestly roles. Chielo's actions and voice are beyond natural human limitations. Her disappearance 'through a hole hardly big enough to pass a hen' (79) when she reaches the abode of the Oracle, while still carrying Ezinma on her back, beats the imagination. She simply dematerialises with the child on her back and becomes present in the cave. Did she have the power to also imbue Ezinma with the same power or is the mere fact that her physical form is in contact with the child enough to also cause the child to dematerialise with her? Unequivocally, Chielo's swift entrance into the cave of the Oracle through a small hole does not mime external reality. Also, her seeming supersonic race through the nine villages with Ezinma on her back is of super-human proportion so much so that Ekwefi could not keep up with her speed. While carrying out her priestly duties, Chielo seems to simply shed off much of her humanity and transforms into a spirit. Hence, the terrified Ekwefi understands that 'there was no humanity' (75) in Chielo's voice, underscoring the fact that a transformation has taken place. The Priestess has been bestowed with extraordinary powers while on her assignment for the Oracle, thus, she becomes, largely, a 'spirit being' just like the deity she serves. Although these speculative threads might seem like an 'alternate reality', they are part of a different kind of reality that are ordinary events to Umuofia people who understand the complexity of Chielo's role as a priestess.

Another fascinating speculative element in *Things Fall Apart* pivots around the activities of the *egwugwu* in chapter ten of the novel. Achebe takes the reader to communal ceremonies where masked spirits emerge from ant holes in the ground. They make 'a terrifying spectacle', (65) with their unsettling macabre sounds of '*Aru oyim de de de dei!*'which hover around the dark, closed hut like tongues of fire. The ancestral spirits of the clan are now abroad and are ushered into the village square as '[t] he metal gong beat continuously now and the flute, shrill and powerful, floated on the chaos' (65). The extraordinary is presented as if it were commonplace, and in this manner, magical elements blend with the real world of Umuofia. The idea of ancestral spirits who live in the earth and emerge from their underground home for the ceremonies in the clan, is one of the remarkable speculative threads in *Things Fall Apart*.

> The drum sounded again and the flute blew. The egwugwu house was now a pandemonium of quivering voices: aru oyim de de de dei! Filled the air as the spirits of the ancestors, just emerged from the earth, greeted themselves in their esoteric language. The egwugwu house into which they emerged faced the forest, away from the crowd, who saw only its back with the many-coloured patterns and drawings done by specially chosen women at regular intervals. These women never saw the inside of the hut. No woman ever did. (65)

When the nine *egwugwu* who represent each of the Villages in Umuofia finally emerge, their entrance to the village square is heralded by their cryptic greetings to one another in guttural voices. 'A steady cloud of smoke rose from the head of Evil Forest' (67), their leader who boasts aloud, 'I am Evil Forest, I kill a man on the day that his life is sweetest to him ... I am Evil Forest, I am Dry-meat-that-fills-the-mouth, I am Fire-that-burns-without-faggots' (68). The *egwugwu* are ancestral spirits that act as protectors and mediators in the world of the living. They emerge from under the ground to intervene in difficult cases in Umuofia and their decision on any case is sacrosanct. This further goes to accentuate the mediatory roles of ancestors who, although are dead, continue to exit and re-enter the community. They help maintain order and remain relevant to the world of the living.

The weeping of the Mother of the Spirits of the land is a remarkable speculative thread in the novel. During the annual worship of the earth goddess, Umuofia is thrown into a great pandemonium when Enoch tears off the mask of an *egwugwu* for flogging him. The implication is that Enoch has not only desecrated an *egwugwu* but has killed an ancestral spirit. The omniscient narrator recounts:

> That night the Mother of the Spirits walked the length and breadth of the clan, weeping for her murdered son. It was a terrible night. Not even the oldest man in Umuofia had ever heard such strange and fearful sound, and it was never heard again. It seemed as if the very soul of the tribe wept for a great evil that was coming – its own death. (136)

Do spirits really die? Generally, the idea of the death of a spirit being might not seem comprehensible. Thus, Svec and Winiski maintain that the enduring debate over what qualifies as science fiction and speculative fiction revolves around 'the plausible and the possible' (37). However, for members of Igbo community of Umuofia, the cries of the Mother of the Spirits of the land at the killing (unmasking) of one of her sons (masquerade), is not only conceivable but tangible. This represents not just an alternate or parallel universe, but a different reality that coexists with their real world. Achebe's portrayal of these two worlds as one, speaks to the complexities of the worldview of Umuofia community where, similar to humans, spirits cry and mourn the loss of their children and loved ones.

Unlike in *Things Fall Apart*, where Agbala and her priestess, Chielo, are minor characters, in *Arrow of God*, the great deity – Ulu and his chief priest, Ezeulu, are the protagonists. In the second chapter of the novel, the author fittingly acquaints the reader with the interesting background that necessitates the installation of Ulu in the community as well as the investiture of his, now larger-than-life, priest, Ezeulu. The following excerpt points to the remarkable folkloric feature and the role of the supernatural in the genesis of the community that is now known as Umuaro:

> In the very distant past, when lizards were still few and far between, the six villages – Umuachala, Umunneora, Umuagu, Umuezeani, Umuogwugwu and Umuisiuzo – lived as different peoples, and each worshipped its own deity. Then the hired soldiers of Abame used to strike in the dead of night, set fire to the houses and carry men, women and children into slavery. Things were so bad for the six villages that their leaders came together to save themselves. They hired a strong team of medicine-men to install a common deity for them. This deity which the fathers of the six villages made was called Ulu. Half of the medicine was buried at the place which became Nkwo market and the other half thrown into the stream which became Mili Ulu. The six villages then took the name of Umuaro, and the priest of Ulu became their Chief Priest. From that day they were never again beaten by an enemy. (14–15)

From this extract, it is clear that the potency of Ulu is not questionable since after his installation, the hitherto incessant attacks from neighbouring villages cease and there is peace within the six villages. The chief priest of Ulu, Ezeulu is reputed to be half man, 'the other half *mmo* [spirit] – the half that was painted over with white chalk at important religious moments. And half of the things he ever did were done by his spirit side' (192). The priest mediates between the people of Umuaro and the Ulu deity and plays decisive roles in the community's two most important religious celebrations; the festival of the Pumpkin Leaves and that of the New Yam. For a chief priest who has both human and spirit sides, Ezeulu is not an ordinary man – he has the power to seamlessly traverse the human world and the world of the spirits.

In the Umuaro world of *Arrow of God*, Achebe also brings the mystery of the ancestral 'night spirits' (225) to the fore. These are spirits who are beyond human knowledge and roam the village square during special communal ceremonies. Thus, Umuaro is a community that is spiritually guided by the magical character of Ezeulu and his famous deity. There appears to be a kind of mythopoeia in the narrative. Mythopoeia or myth-making is the creation of artificial mythology which, in fictional narratives, refers to the depiction of the real world as having an undercurrent of magic or fantasy. For example, the village of Umuachala in Umuaro normally celebrates a minor feast of *Akwu Aro*. This is a memorial offering of foofoo and palm nut soup by widows to their departed husbands. On the night of the feast, every widow keeps the food outside her hut. In the morning the bowls become empty because her husband comes up from *Ani-Mmo* (the spiritual world) and eats the food (194).

Like Umuofor in *Things Fall apart*, Umuaro is a community where there is a fluid boundary between the human world and the world of the spirits. The burials of accomplished men are heralded by the *ogbazuluobodo* ritual where a human transforms into a spirit and guides the departed to the ancestral world. This boundary-fluidity becomes more evident when, as soon as Obika – Ezeulu's son, is geared up in *ogbazuluobodo* costume and staff, he transforms into a spiritual being and cries 'Ewo okuo! Ewo okuo!' (255). The drummer who has been talking and laughing with him before this time, throws down his sticks and hastily blows out the offending light because spirits abhor lights. Obika, now a spirit, plants the *ogbazuluobodo* staff into the earth and it reverberates. Like a spirit, indeed, he pulls it out again and vanishes like the wind in the direction of Nkwo market. In this instance, the *ogbazuluobodo* staff becomes a transformative symbol; and Obika's contact with it ushers him into the world of the spirits which enables him to accomplish the funeral night-ritual of running through all the clans of Umuaro in astonishing speed.

Ezeulu is a chief priest who seems to relish the fact that the deity sometimes uses him as a whip (208) on members of Umuaro. Although he is half-spirit, half-human, his spiritual side is often emphasized. Through the central role he plays in his capacity as chief priest of Ulu during the Feast of the New Yam (201), Ezeulu uses the potency of Ulu to exact vengeance on his detractors. However, Ulu – the Oracle he serves – appears to be a war-monger-deity. Ulu, clearly warns him that the war he is meddling with is a war of the gods and not of men. The deity speaks to his chief priest in a human voice and when the priest appears headstrong to pursue his personal agenda, Ulu swiftly calls him to order like an errant child:

> 'Ta! Nwanu!' barked Ulu in his ear, as a spirit would in the ear of an impertinent human child. 'Who told you that this was your own fight?' Ezeulu trembled and said nothing, his gaze lowered to the floor. 'I say who told you that this was your own fight to arrange the way it suits you? You want to save your friends who brought you palm wine he he-he-he-he!' Only the insane could sometimes approach the menace – a dry, skeletal laugh and mockery in the laughter of deities. 'Beware you do not come between me and my victim or you may receive blows not meant for you! Do you not know what happens when two elephants fight? Go home and sleep and leave me to settle my quarrel with Idemili, whose envy seeks to destroy me that his python may again come to power. Now, you tell me how it concerns you. I say go home and sleep. As for me and Idemili we shall fight to the finish; and whoever throws the other down will strip him of his anklet.' (191–192)

The 'barking' voice of Ezeulu's deity in *Arrow of God*, contrasts markedly with the conduct of Chielo's Agbala, the Oracle of the Hills and the Caves, in *Things Fall Apart*. In Umuofia, the power of Agbala is felt through the deity's interactions with Chielo, especially, through her messages to the community and the power Chielo wields as a priestess. There seems to be a comfortable distance which exists between the community members and the Oracle that is perceptible to the reader. On the other hand, the interaction of the deity of Umuaro, Ulu, and his priest, Ezeulu, is presented in a manner that clearly indicates that Ulu is from another world – a spirit that has made his abode among the members of the Umuaro community. His startling 'dry, skeletal laugh[ter] and mockery' (191) which are directed at his priest projects him as a rash and malevolent deity whose words to Ezeulu reveals his belligerent nature. For instance, he cautions Ezeulu, 'I say who told you that this was your own fight to arrange the way it suits you? … Beware you do not come between me and my victim or you may receive blows not meant for you!' (191). Ulu seems to have made good his threat with Ezeulu's eventual madness and the sudden untimely death of his favourite son.

Achebe seems to weave the speculative into the realistic socio-cultural dynamics of Umuofia and Umuaro, in his *Things Fall Apart* and *Arrow of God*. The relationship between Ulu and his chief priest, Ezeulu, is shaped by the personality of the deity who is perceived as powerful, rash and egocentric while, on the other hand, the relationship between Agbala and her priestess, Chielo, is also guided by the Oracle's personality which comes across as powerful, temperate, and considerate. The different traits of Agbala and Ulu not only influence the relationships between the deities and their priests but also shape the different views that members of Umuofia and Umuaro hold of these deities and their 'half-spirit half-human' priests. These views and relationships further deepen the speculative-contents of the narratives. Interestingly, what is generally considered as the speculative in fictional writings might continue to evolve as the conversation around what should be defined as speculative fiction continues to expand. Notably, Robert Scholes recognizes this genre as a 'fictional form that is both old and new, rooted in the past but distinctly modern, oriented to the future but not bounded by it (*Structural Fabulation*: 27)'. *Things Fall Apart* and *Arrow of God* fit this frame – they feature speculative elements that are firmly grounded in realistic sceneries with life-like characters in Umuofia and Umuaro communities.

CONCLUSION

Evidently, Achebe's *Things Fall Apart* and *Arrow of God* offer productive sites for the exploration of literary documents which showcase a considerable mix of speculative stories. Thus, a re-reading of the author's canonical first novel, *Things Fall Apart*, and his third work, *Arrow of God*, placed within a speculative-narrative-sensitive-lens, reveals substantive snippets of what Svec and Winiski describe as a 'more modern speculative fiction' which offers a 'fascinating mix of science, personal experience and fantasy' (43). This genre of contemporary speculative fiction describes novels that are inspired by mythology, folklore, and fairytales that combine imagined technologies with elements of magical realism. Achebe's *Things Fall Apart* and *Arrow of God*, arguably, suit this category when viewed from the angle of mythology and folklore. Although the two novels represent elements of the speculative, they are set in real and identifiable world and, thus, may be appropriately labelled as having pockets of speculative stories within their narratives. A re-reading of these texts through a speculative lens provides an alternative interpretation which gives more insight into their socio-cultural settings.

WORKS CITED

Achebe, Chinua. *Arrow of God*. London: Heinemann, 1964.

—— *Things Fall Apart*. London: Heinemann, 1958.

Atwood, Margaret. *In Other Worlds: SF and the Human Imagination*. New York: Nan A. Talese, 2011.

Doubinsky, Sébastien. 'Jordan Krall's Speculative Fiction'. *Alluvium* Vol. 6, No. 1, 2017.

Kincaid, Paul. 'Against a Definition of Science Fiction'. *World Literature Today*, SF Issue Vol. 84, No. 3, 2010.

Merril, Judith. 'Introduction'. In *SF: The Best of the Best*, Judith Merril (ed.). New York: Delacorte Press, 1967.

Scholes, Robert. *Structural Fabulation: An Essay on the Fiction of the Future*. Notre Dame, IN: University of Notre Dame Press, 1975.

Svec, Michael and Mike Winiski. 'SF and Speculative Novel: Confronting the Science and the Fiction'. In *Science Fiction and Speculative Fiction: Challenging the Genres*, Paul L. Thomas (ed.). Rotterdam: Sense, 2013: 35–57.

Waggoner, Diana. *The Hills of Faraway: A Guide to Fantasy*. New York: Atheneum, 1978.

Contemporary Ugandan Speculative Fiction

A Passing Fad or an Emerging Canon?

EDGAR NABUTANYI

INTRODUCTION

Helon Habila declares that *The Granta Book of African Short Stories* should be canonized if after ten years of its publication it still 'illuminates the preoccupations and concerns, literary and social, of the times' (10). Although his test of time as a measure of literary merit is vastly shorter than the standard 100 years, I agree that thematic relevance and stylistic innovativeness are robust grounds for canonization. Using Habila's yardstick of canonization, I strongly recommend for the inclusion of Innocent Acan's 'The Machodugo', Lillian Aujo's 'Where Pumpkin Leaves Dwell', and Dilman Dila's 'The Last Storyteller' in the Ugandan literary canon. My justification for their inclusion is due to how their respective authors deploy tropes of speculative fiction to showcase Ugandan dystopic social collectives. Furthermore, I argue that the speculative fiction motifs of the selected texts enable their writers to become Wale Adebanwi's 'social thinkers'. Adebanwi argues that writers are 'not merely intellectuals whose work mirror or can be used to mirror social thought, but are social thinkers themselves who engage with the nature of existence and questions of knowledge' ('The Writer as Social Thinker'; 406). The essence of Adebanwi's argument is that writers do more than faithfully imitate the reality of their societies. They use their work to discuss and take a stand on pertinent issues affecting society. African writers may not tell readers how to think about important societal issues, but their work raises consciousness concerning what issues are worth thinking about.

The importance of writers in distilling profound insights about societal conditions is a point of agreement between Adebanwi and Habila. It is this role of writers that justifies the inclusion of any author's works in the canon of his/her society. My proposition to include Acan's 'The Machodugo', Aujo's 'Where Pumpkin Leaves

Dwell', and Dila's 'The Last Storyteller' in the Ugandan literary canon recalls Janice Radway's argument that 'works are selected on the basis of aesthetic achievement' (*Reading the Romance*: 3). She goes on to note that canonized texts may not 'necessarily be representative of the large section of the population that had never read them' (3). In these quotations, Radway makes two important comments. First, that canons are decided on aesthetic value of the texts. Second, that canonization is an elitist decision by elite institutions such as the media, literary clubs, universities, and libraries. Radway is right on insisting that aesthetic excellence is a mark of canonization. This means that elite institutions are central in canon formation. However, this model has been disrupted by online platforms of consuming literature and bestowing literary merit.

Cyber platforms in curating, circulating, and awarding literary distinction have introduced a new model of canonization. This new method of sanctification of literary texts has a place for speculative short fiction. As such, Acan's 'The Machodugo', Aujo's 'Where Pumpkin Leaves Dwell', and Dila's 'The Last Storyteller', which are published, circulated and ratified by online literary collectives, can claim canonization in the Ugandan literary public sphere. My argument recalls John Guillory's differentiation between the 'traditionalists and relativists' perspectives of canon formation ('Canon' quoted in Grishakova 'Poetics of the Return': 23). Marina Grishakova explicates Guillory's postulation by arguing that while 'the traditionalist aesthetic theory of canon formation is based on the underlying notions of value and quality' (23), a relativist view argues that 'the canon is only a cultural artefact as a matter of social agreement' (24). Grishakova's argument underscores the point that canon formation is a compromise undertaking. While it demands for the recognition of 'a corpus of classical texts', (Grishakova 25) it also acknowledges societal involvement in the decision of what represents their aesthetic and thematic sensibilities.

I apply Grishakova's postulation to Acan's, Aujo's, and Dila's short stories to justify their claim for inclusion in the Ugandan canon, as their canonization can be negotiated on aesthetic fineness and pedagogical robustness. All three stories have been published and circulated by an important online platform, Heady Mix, under the title *Afrofuturism*, echoing a distinctively well-recognized African-American sub-genre that addresses African-American themes and 'concerns in the context of twentieth century techno-culture' (Cruickshank 'Introduction': 1). Nonetheless, the editor of this collection borrows and adopts the term to reflect African and African diasporic literary experiences or 'stories that take place in the future [about] Africans living in Africa' (quoted in Cruickshank 1–2). These textual markers gesture to the potential for the canonization of African speculative fiction. I posit that Acan's, Aujo's, and Dila's short stories, which describe the lives of Ugandans living in Uganda, can claim for inclusion in the Ugandan canon. The aforementioned argument is reinforced by the global significance attached to online platforms like the Jalada Prize in adjudicating the value of African literary texts. The circulation of the selected texts via the previously cited literary collectives complicates Trevor Ross's argument that 'canons are made and preserved within critical and academic institutions as well as cultural establishments such as public libraries, publishing houses' (4–5) in two paradoxically interesting ways. First, the 2019 online publication of Acan's, Aujo's, and Dila's short stories by Heady Mix underlines an emerging literary tradition. Second, the emerging tradition cited above is yet to be incorporated in Uganda's literary, critical, and academic institutions. Nonetheless, I posit that the selected Ugandan speculative fiction stories published and circulated by virtual literary collectives has assumed a mark of excellence that places them in the Ugandan literary canon.

UGANDAN SPECULATIVE FICTION: AN EXPLORATION

There is a paucity of critical works on Ugandan speculative fiction for two reasons. First, unlike Nigeria and South Africa, Uganda does not have a long and established tradition of speculative fiction writing. In fact, apart from Austin Bukenya's *A Hole in the Sky*, there have been hardly any speculative texts published until that of Acan's 'The Machodugo', Aujo's 'Where Pumpkin Leaves Dwell', and Dila's 'The Last Storyteller'. Second, the Ugandan literary canon is skewed towards verisimilitude fiction in the three major genres of the nationalist novel, liberational drama, and consciousness arousing poetry. This is further apparent in the absence of speculative fiction on the Ugandan secondary school and university literature curriculum, graduate academic research, and in literary reviews in Ugandan newspapers.[1]

Given that the short story generally, and speculative short fiction specifically, exists at the periphery of the Ugandan canon and literary scholarship, the circulation of Acan's 'The Machodugo', Aujo's 'Where Pumpkin Leaves Dwell' and Dila's 'The Last Storyteller' via the internet and their authors' gaining recognition and winning international prizes by reputable literary curating institutions like The Caine Prize for African Writing, suggests a shift in the parameters of literary canon formation in the country. I aver that the availability of information about Acan's 'The Machodugo', Aujo's 'Where Pumpkin Leaves Dwell', and Dila's 'The Last Storyteller', through internet curated and circulated journalistic works and blog entries, gestures to the subversion of the Ugandan canon, even when such publications focus mainly on these authors' biographical information.

The case in point is Nyana Kakoma's *Sooo Many Stories* website which notes that Acan is a 'Ugandan speculative fiction writer' and a winner of 'the *Writivism* Short Story Prize in 2016 and [that she] has been published by *AFREADA Magazine*, *Omenana Magazine*, and *Brittle Paper*'. Similarly, a website called *LéO Africa Review* notes that 'her most recent work is a children's book titled: *The Pearl Trotters in Black, Yellow, Red*', whose 'protagonist Nyamayarwo negotiates with death and her five Sentinels'. While the above websites do not provide us with incisive analyses of her work, they help us get a glimpse of the themes and techniques that characterize her work. In foregrounding the prestigious accolades that she has won and the avenues where she has published her work, these websites identify her as an important Ugandan speculative fiction writer.

There is a similar dearth of information about Aujo. It is only the website: *Whats On Africa* and a blog by James Murua that provide us with mainly biographic information about Aujo. While *Whats On Africa* informs us that 'she was the winner of the inaugural Jalada Prize for short fiction', the blogger James Murua reports that she won the Jalada Prize with her short story 'Where Pumpkins Dwell ... which is about a young girl who is left by her mother in a village as the mum gets an education and the adventures she has'. The information supplied by the aforementioned online publications underline the excellence of Aujo's work by referring to the prizes she has won and the prestigious literary collectives such as The Caine Prize for African Writing, Jalada, Writivism, and Femrite that have published her work. However, in her interview with Geoff Ryman in *Strange Horizons*, Aujo offers us insights into the importance of her work in reimagining the Ugandan literary canon. She says that she 'was not thinking about genre' when she wrote 'Where Pumpkins Dwell', but that her current 'writing [is] more consistently thinking about it [genre] more consciously'. She also states that her 'first long work will be Afro Sci-Fi ... I have a plan. I've written a synopsis ... I'm actually thinking I can take the first 10,000 words as a novella and submit it to Ivor Hartman's Afro SF series'.

The trajectory of her work and the attention that it has attracted from arbiters of aesthetic excellence and thematic robustness in speculative fiction such as Geoff Ryman and Ivor Hartman underscore her contribution to the field of speculative fiction in Uganda.

The potential for Acan and Aujo to disrupt the Ugandan literary canon has perhaps been built on the work of Uganda's pioneering speculative fiction writer – Dilman Dila. This is because his work has attracted significant critical attention, in part due to his position as the most prolific Ugandan speculative fiction writer to date. He has published an anthology of speculative short fiction entitled *A Killing in the Sun*, and has also had his work published in respected publications such as the *New Orleans Review* in its edition The African Literary Hustle. His work has been flatteringly described by Mukoma wa Ngũgĩ and Laura Murphy as having 'outgrown African literature's obsession with the realist novel in English' ('This Hustle is Not Your Grandpa's African Lit.': 2). The peer recognition of his unique voice and style underlines Dila's greatness as Uganda's leading speculative short fiction writer.

Dila's prominence as a speculative fiction author is also underlined by Nedine Moonsamy's and Timothy Sean Wright's reading of *A Killing in the Sun* (2014). Moonsamy argues that *A Killing in the Sun* is an 'imaginative and poetic rendition of African futures where other-worldly technology blends in innovative ways with contemporary customs, culture, and concerns' ('Aliens and Insecticide': 76). She also notes that Dila grapples with 'the realities of the neoliberal African state' (75) and that he also illuminates 'various paradoxes around the consumption of internationally sponsored insecticide and its subsequent cost on local societies' (76). Similarly, Timothy Sean Wright also foregrounds the brilliance of Dila's stylistic engagement with the theme of environmental disasters in 'Third World' countries like Uganda. He argues that Dila's short story 'emerges at the confluence of ecological catastrophe and corporate neocolonialism' ('Surviving the African Anthropocene': 142). In the above passages, Moonsamy and Wright underline two major aspects of Dila's short stories. First, they agree on the point that his short stories grapple with important societal issues. Second, they concur that he deploys sci-fi tropes to imagine Ugandans living in a country plagued with ecological catastrophes. I argue that the aesthetic excellence and thematic relevance of Dila's work has disrupted the Ugandan literary mark of excellence.

UGANDAN SPECULATIVE FICTION: A PASSING FAD OR AN EMERGING CANON

Acan's, Aujo's, and Dila's speculative short fiction constitute what Sofia Samalar ('Towards a Planetary History of Afrofuturism'), Lizelle Bisschoff ('African Cyborgs'), and Moradewun Adejunmobi ('Introduction: African Science Fiction') have labelled an African speculative fiction writing tradition in the continental literary public sphere. While Samalar notes that 'in the twenty-first century, African science fiction has become visible', (175) Bisschoff reports the emergence of 'a genre of speculative fiction dealing in imaginary and imaginative concepts often related to science and technology' (4). In the above passages, the scholars underline two important points. First, that African sci-fi is an emerging continental literary genre. Second, that African sci-fi exhibits traits of this genre advanced by Darko Suvin ('On the Poetics of the Science Fiction Genre') and Istvan Csicsery-Ronay ('The SF of Theory'). Suvin argues that sci-fi is a 'literary genre whose main formal device is an imaginative framework alternative to the author's empirical environment' (375). Similarly, Csicsery-Ronay notes that a speculative fiction text lies in the 'gap … between the conceivability of future transformation and the possibility of their actualisation' (387). I apply the two definitions of speculative fiction as a genre that imaginatively depicts a future that is inconceivable in the present to Acan's 'The Machodugo', Aujo's 'Where Pumpkin Leaves Dwell', and Dila's 'The Last Storyteller' because they simultaneously forecast the future and retrospectively imagine a Ugandan dystopic past. They also offer 'an alternative [insight into] the empirical environment' of Uganda (Adejunmobi: 267).

Adejunmobi uses *District 9* (Blomkamp, 2009) to illustrate how African sci-fi 'allegorises [a post]colonial encounter between Africa and a globalised Western world' ('Introduction': 266). She goes on to argue that it is an example of 'how African writers use speculative fiction to interrogate the socio-political arrangements that inform the African condition in the current age' (265). The essence of Adejunmobi's reading of *District 9* as postcolonial allegory is to show how speculative fiction can allow readers to engage with important issues affecting society. Her reading also suggests that speculative fiction writers use their fiction to imagine techno-environmental disasters waiting to happen such as climate change that is devastating some parts of Africa. I foreground Adejunmobi's argument in my reading of Acan's, Aujo's, and Dila's sci-fi short stories to argue that Ugandan speculative fiction offers us a vision of the impending catastrophic future. I start by exploring the catastrophic intersection between modernity and tradition in Acan's 'The Machodugo'. Next, I read Aujo's 'Where Pumpkin Leaves Dwell' as phantasmagorical re-enactment of Uganda's trauma with HIV/AIDS. Lastly, I read Dila's 'The Last Storyteller' as an exploration of the coping mechanisms open to those who are traumatized by environmental disasters.

Innocent Acan, 'The Machodugo'

'The Machodugo' adopts a folkloristic motif to weave a cautionary tale of the tragic consequence that arises when modernity comes in contact with tradition. It is structured as a tragic love triangle, whose vertices are the Machodugo of Ondravu called Aasi, the city girl Philomena Andrua and the Moon goddess. The structure of the short story alludes to Okot p'Bitek's *Song of Lawino* and its critique of cultural imperialism symbolized by Philomena (modernity) and the Moon goddess (tradition). Thus, the text is a cautionary metaphor for the traumatic outcomes in the conflict between modernity and tradition. The conflict of the short story starts when Aasi, who is entranced with Philomena's beauty, abdicates his duty of providing clean sweet water to the village of Ondravu. He dedicates himself to the service of Philomena, forgetting that he serves at the pleasure of the Moon goddess. The omniscient narrator notes that Aasi is turned into a slave to 'the city princess' (22) with devastating consequences for both Aasi and the village. Aasi's infatuation with Philomena distracts him from his role as a provider of clean sweet water to Ondravu. The narrator observes that when the well becomes unkempt, his sister Amaru travels to Ondravu to warn him to change his ways lest he faces the wrath of the Moon goddess (23). The anticipated change is short lived because when Philomena angrily complains, 'I hear you're now giving my clear sweet water to the villagers – you unfaithful well … I don't feel special anymore' (23), Aasi's resolve to serve the people of Ondravu crumbles. Torn between serving 'his love who was angry with him' (23) and the people of Ondravu he decides to 'serve Philo alone [and as a result, the well in] Odravu spat up green bitter water for all but one girl. The Moon goddess was furious. A machodugo dared to love another over her' (23). The Moon goddess's punishment is swift and devastating because when next the girls go to the well, they discover that 'Philo's well had run dry' (24).

Acan reconfigures tropes from speculative fiction to help reimagine the theme of a cultural clash, as Acan's literary agenda 'entails a change of the whole universe of the tale, or at least crucially important aspects thereof' (Suvin, *Metamorphoses of Science Fiction*: 64) to showcase the trauma of cultural clash. She utilizes the folklorist cautionary tale motif, allusion, allegory, and symbolism to weave a convincing tale of the horrific impact of cultural conflict on the fictional village of Ondruva. The short story echoes p'Bitek's *Song of Lawino* and its postcolonial epistemes, which is eloquently crafted in the character symbolism of the text. While Philomena – the beautiful city girl – signifies worldly modernity, the Moon goddess symbolizes traditional sanctity. Aasi's struggle to serve either Philomena or the Moon goddess is the quintessential struggle between an amoral modernity and an idyllic traditional past. It is important to note that it is the village which pays the price represented by a well that dries up. The village's loss is further stressed by Philomena's nonchalant comment that her 'father [was] going to put piped water in [their] house' (24). This passage underscores how oblivious Philomena is to the disruptions she has caused. Her characterization as a quintessential anima is comparable to the Biblical Delilah who causes Samson's tragedy. It also underlines two important points about the conflict of the short story. First, it suggests that modernity wins. Second, it underscores the disruptive nature of the cultural clash. This is because when Philomena, the symbol of modernity, arrives in Ondruva, the village's way of life is turned upside down as symbolized by the dried up well. When the well (which signifies the communal spirit of the village because it brings the village's young women together) dries up, Philomena declares that her father will bring piped water. Her declaration symbolizes the death of the village. Comparable to Aasi who is 'shattered in nothingness', (24) the destruction of the village recalls Obierika's profound statement in Achebe's *Things Fall Apart* that the White man had put a knife of the things that held them together and that they had now fallen apart (124–125). In a way, Acan shows how the intrusion of modernity into the communal and idyllic life of Ondruva tragically results into the symbolic death of the village. Acan's intertextual mirroring of Achebe, the Bible and p'Bitek reminds us of Guillory's argument that canonical works 'maintain the cultural capital' of a given society (xiii). Acan builds on Achebe's and p'Bitek's seminal works to underline the catastrophic consequences of cultural clash on the depicted society.

LILLIAN AUJO, 'WHERE PUMPKIN LEAVES DWELL'

I now turn to how 'Where Pumpkin Leaves Dwell' deploys a phantasmagorical motif to portray the trauma of HIV/AIDS. Aujo joins a cohort of Ugandan artists who have used their art to debate the impact of HIV/AIDS on Ugandan society.[2] The short story is set in a quintessentially Ugandan village and features a typical Ugandan practice where parents send children to their grandparents in the village. The separation between the prodigious Pat and her mother takes place because the city 'is for adults and there are no children allowed … because there are no children there' (27). This sets Pat on a quest to understand the village, its inhabitants, and way of life. She discovers that the village is called Lyakanshunsha because 'pumpkins grow everywhere in the village' (29). However, the real revelations are the tragic secrets of the community, symbolized by the graves in Bwengye's homestead. In a feat of delirium that her grandparents blame on a fever: 'you are sick Pat. It was just a bad dream, and you have malaria' (33), Pat discloses how HIV/AIDS has devastated the village. This is underlined in the passage: 'the mould of soil [that] will sink and gape', 'the dead parent – one has breasts and the other a flat chest' (32), and Bwengye's angry declaration 'have you forgotten about the graves in our backyard? And all these orphans your children left behind? (33). These passages unveil the tragic loss that the village has had to endure.

The horror of HIV/AIDS that Aujo grapples with remind us of Suvin's argument that 'SF can finally be written only between the utopian and an anti-utopian horizon' (76). Suvin underlines the paradox of sci-fi, namely, that while scientific advancements are supposed to usher in happiness, they often instantiate suffering. Aujo's 'Where Pumpkin Leaves Dwell' uses the aforementioned sci-fi trope to explore the anti-utopian horizon of Ugandan society in the face of the HIV/AIDS pandemic. Thus, as Adejunmobi notes, the text 'enforce[s] a strict separation between works incorporating supernatural or magical elements and those exploring the outcomes of speculative technologies [that have] a long history in sf studies' (269). Adejunmobi underlines two key tropes of speculative fiction, namely, the focus on the supernatural/magical and the foregrounding of speculative technologies in the texts. One of Adejunmobi's dual characteristics of speculative fiction – the textual focus on the supernatural – is evident in Aujo's short story. Aujo deploys magical elements to 'interrogate the socio-political arrangement that inform the African condition in the current age' characterized with HIV/AIDS (265). The short story's dystopic social reality is symbolized by the numerous graves in Bwengye's homestead, the many orphans this family has to care for and the resultant stress that manifests as domestic violence and belief in the supernatural. It is these horrors of a society turned upside down that Aujo unveils by using her protagonist's delirium.

Adejunmobi's argument that speculative fiction is interested in representing postcolonial reality (265) is apparent in Aujo's problematizing of the urban-rural journey trope, which has been eloquently delineated in canonical African postcolonial fiction such as Alan Paton's *Cry, the Beloved Country*, Ngũgĩ wa Thiong'o's *Petals of Blood* and Okot p'Bitek's *White Teeth*. Aujo uses Pat's relocation to Lyakanshunsha to spotlight the horrific impact of the HIV/AIDS epidemic on the people; the presence of graves that trigger Pat's delirium symbolize the devastation of rural Uganda in the 1980s. The depiction of the devastation allows Aujo to subvert the rural-urban motif in postcolonial African fiction, in which the rural area is constructed as a site for rejuvenation. While the rural areas have been invested with therapeutical powers in postcolonial African fiction, Aujo destabilizes this motif to accentuate the prevalence of HIV/AIDS induced suffering in the depicted society. The speculative fiction credentials of this story are underlined by the improbability of Pat's knowing, let alone disclosing the secrets of the village. This supernatural power is unveiled in the passage: 'but how can she know all that when she wasn't even born' (34). The grandparents' surprise at Pat's mystical knowledge of the village's catastrophic past is one of the many speculative characteristics of the short story. This reminds us of Benjamin Christopher Manadue's and Karen Diane Cheer's argument that speculative fiction is 'driven by the need to provide some sort of manageable interpretation of an increasingly complex and unstable social and intellectual reality' ('Human Culture and Science Fiction': 1–2). Manadue and Cheer are right when they argue that speculative fiction allows writers to imagine a complex and unstable world, and this is apparent in Aujo's speculative framing of Pat's supernatural powers to engage with the complex matter of the devastation of HIV/AIDS.

In 'Where Pumpkin Leaves Dwell', intertextual allusions are made to the Greco-Judean trope of the Garden of Eden through references to the pumpkin vines and the serpent. Thus, Pat's cry that 'the snake and the pumpkin vines want to swallow me' (33) simultaneously alludes to entanglement (pumpkin vines) and paradise lost in Lyakanshunsha (snake). Lyakanshunsha's paradise lost is scripted in the graves of Bwengye's children, and the enormity of this catastrophe is articulated in phrases like 'all these orphans', and 'your children'.

The two phrases cohere with the verb 'left' to show that all Bwengye's children that Pat sees in her visions are dead. Pat's visions symbolize the stylistic aspects of speculative short fiction that enable the writer to disclose the enormous loss of life to HIV/AIDS. Relatedly, Bwengye's wife's taking on Pentecostal Christianity symboliszs one of the coping mechanism Ugandans adopted to deal with the overwhelming cost of HIV/AIDS at the time. Unable to make sense of the epidemic, many Ugandans resorted to superstition, including Pentecostalism as a coping mechanism for the horror they were witnessing.

DILMAN DILA, 'THE LAST STORYTELLER'

Lastly, I turn to how Dila deploys speculative fiction tropes and storytelling to explore the means of surviving the trauma of post-environmental disasters in 'The Last Storyteller'. This short story is set in post-apocalypse Uganda and it follows the trials and tribulations of a surviving human storyteller called Aya. The short story is structured as Aya's quest to unearth a past she has obviously repressed in her unconsciousness. The quest is triggerred by a stranger who seeks her out after reading her story and claims that he is her husband. Aya first goes to the dead lake that had inspired her story, hoping that the sight of the lake would jog her memory. After the visit to the dead lake, she logs into her own story hoping she can glean information about 'Obaraf – a name given to a person born on a day that ice does not melt at sunrise' (98). Her quest to divulge the identity of a man, who claims to be her husband, ends at the Population Records Office of BDR (Department of Birth and Death Registration). Her search for a husband she might have erased out of her memory at the BDR brings back the statement: 'this person has never existed' (101). This leads to her conclusion that the purported husband is a 'real-life print that bled real blood and had the scent of a real person. Maybe the machine had become so efficient that it could print a real-life person' (101). She also speculates that the man who claims to be her husband has been sent by either her mother to lure her into a conversation about the past or IT to make her happy so that she can create human veepics that people are hungry for (97, 102). The short story ends with Aya deciding to leave Fish city because 'she would not go back home. Not to the ghost that waited for her' (103).

'The Last Storyteller' echoes Dila's confession that his stories 'go beyond the edge of reality and that it is not impossible for those events to happen' (Kakoma 'Dilman Dila on Being an All-Round Storyteller').

It also reminds us of Moonsamy's observation that 'representations of eco-apocalypses, where all humanity perishes because of our collective actions, stresses the need to move beyond the political into more ecoambiguous terrain' ('Aliens and Insecticide': 78). While we are not told the cause of the horrific dystopia that Dila chronicles in the short story, the dead environment, the mechanized human relationships, and loneliness underline societal dystopia. Ironically, in a society that is so scientifically advanced that it can make human beings called 'prints', there is pervasive unhappiness and loneliness. It is Aya's speculation that IT could have created an illusion of a living person that underlines the paradox at the heart of speculative fiction. This is the idea that technological advancement does not translate into happiness. This perhaps explains Aya's value as the only surviving human storyteller. She notes that IT was aware of her worth as the only surviving human storyteller, and deduces that IT has printed for her a husband to 'ease her loneliness, for IT knows individual parts need to bond with other individual parts to be happy' (102). This, however, recalls a terrifying horror evocative of George Orwell's *1984*'s logic of authoritarian regulation. IT is comparable to the all-seeing, all-knowing Big Brother in Orwell's novel. In both texts, intrusive and reconnaissance technology intrudes into the lives of the protagonists with devastating consequences.

Dila uses stream of consciousness and a 'story within a story' – Aya's awarding winning 'Birds on the Boat' – to showcase the horrors of a totalitarian society on account of environmental destruction. For example, the 'story within a story' technique allows him to contrast an idyllic past before an environmental disaster with the harsh reality of the present ecological catastrophe. This is underlined by a series of rhetorical questions: 'Had birds really filled the skies? Had trees and plants really grown all over the planet with all kinds of animals feeding off them? Had there been a lake with boats and fish' (96)? Dila uses rhetorical questions to underscore the devastation of society by an environmental catastrophe. The dead and unhappy world of the text is further accentuated by the mechanization of the society symbolized by mechanical waitresses and a population that is 'hungry for a human story' (96–97). Dila's writing presents the horror of a post-Armageddon Ugandan future. The physical devastation is accentuated by Aya's stream of consciousness that allows Dila to place readers into her mind as she searches for 'any memory surgeries' (101) and decides against searching for answers from her mother because she 'would want to talk about other things, things that Aya would rather stay buried deep in her subconscious' (100). This reminds us of Cathy Caruth's argument that trauma is 'an unexpected or overwhelming violent event or events that are not fully grasped as they occur' (*Trauma: Explorations in Memory*: 91). Aya exhibits the classic symptoms of someone traumatized, in this case by an environmental apocalypse, through her decision not to unravel the horror she has suppressed in her unconscious.

CONCLUSION

The thesis of this essay is that the aesthetic finesse and thematic relevancy of the selected Ugandan speculative short fiction serves as a justification for their inclusion in the Ugandan literary canon, on the one hand. On the other hand, I have argued that the circulation of these short stories in respected online platforms such as Jalada and the recognition of their writers by institutions such as The Caine Prize of African Writing underline their provisional inclusion into a Ugandan literary canon. My claim is underscored by Andrew Milner's argument that the 'term canon denotes a set of officially recognised sacred books' ('Review': 425). Milner further states that the 'loathed possibility of the construction of alternative and inferior canons is precisely the reason why one forms a canon in the first place' (425). The thesis of Milner's argument is that a canon is created to ensure the purity of a society's officially designated literary excellence. While I agree with Milner that canons are constructed on an exclusion logic – the separation of texts considered to represent literary excellence from the 'others' – I have argued that the circulation of these selected short stories via online platforms and their recognition by literary curating agencies suggests an emergence of a new Ugandan literary canon and institutions of literary canonization.

Granted, Acan's 'The Machodugo', Aujo's 'Where Pumpkin Leaves Dwell', and Dila's 'The Last Storyteller' may take long to appear on a Ugandan Literature Examination Syllabus, get stocked by the Ugandan National Library, or become foundational texts of a University Curriculum – apart from perhaps appearing on the Popular Literature reading lists. However, their disruption of what constitutes literary excellence has earned them a place in the country's canon. This means that Acan, Aujo, and Dila can join p'Bitek, Moses Isegawa, Jennifer Nansubuga Makumbi, and Susan Kiguli in a Ugandan tradition of literary excellence. This is because they have created an alternative space in the country's literary tradition of excellence for themselves.

Nonetheless, my claim is aware of the fact that a canon is a contested reality, on the one hand, and a compromised outcome given the impossibility of agreeing on a single list of texts as representative of a society's tradition of literary excellence, on the other hand. It is this inherent paradox of canon formation that motivates my suggestion for the expansion of the Ugandan literary canon. This is because 'there is no such thing as a literary work or tradition which is valuable in itself … value is a transitive term: it means whatever is valued by certain people in specific situations' (Eagleton, *Literary Theory*: 11). Drawing on Terry Eagleton's two interesting points, it is worth noting that literary value is ephemeral by nature. This means that the value of a piece of writing is not intrinsic in the text, and literary merit is subjective and contextual. If literary merit is transitory, subjective and contextual, then, Acan's, Aujo's, and Dila's speculative fiction that eloquently explores the recurrent theme of trauma in Uganda deserves literary recognition.

NOTES

1 Suffice to note that the poetry of Okot p'Bitek, David Rubadiri, Susan Kiguli, and Timothy Wangusa have been studied and written about more than the short stories of Monica Arac de Nyeko. Similarly, Moses Isegawa's *Abyssinian Chronicles* and *Snakepit*, and Jennifer Nansubuga Makubi's *Kintu* have inspired conference papers, journal articles and panel discussions at both national and international literary meetings.

2 Uganda musician Philly Bongole Lutaaya, dramatist Andrew Benon Kibuuka, and novelists Mary Karooro Okurut and Moses Isegawa have used art to usefully debate the HIV/AIDS pandemic in Uganda.

WORKS CITED

Acan, Innocent Immaculate. 'The Machodugo'. *Afrofuturism*. London: Heady Mix, 2019: 21–26.

Achebe, Chinua. *Things Fall Apart*. London: Heinemann, 1958.

Adebanwi, Wale. 'The Writer as Social Thinker'. *Journal of Contemporary African Studies* Vol. 32, No. 4, 2014: 405–420.

Adejunmobi, Moradewun. 'Introduction: African Science Fiction'. *Cambridge Journal of Postcolonial Literary Inquiry* Vol. 3 No. 3, 2016: 265–272.

Aujo, Lillian. 'Where Pumpkin Leaves Dwell'. *Afrofuturism*. London: Heady Mix, 2019: 27–34.

Bisschoff, Lizelle. 2019. 'African Cyborgs: Female and Feminists in African

Science Fiction'. *Interventions: International Journal of Postcolonial Studies* Vol. 2, No. 5: 1–18.

Blomkamp, Neill (dir.) *District 9*. 112 mins. Sony Pictures, 2009.

Caruth, Cathy. *Trauma: Explorations in Memory*. Baltimore, MD: Johns Hopkins University Press, 1995.

Csicsery-Ronay, Istvan. 'The SF of Theory: Baudrillard and Haraway'. *Science Fiction Studies* Vol. 18, No. 1, 1991: 387–404.

Cruickshank, Justina. 'Introduction'. *Afrofuturism*. London: Heady Mix, 2019: 1–6.

Delan, Bruce. 'Afrofuturism: From the Past to the Living Present'. *UCLA Newsroom*, 3 September 2020. www.newsroom.ucla.edu (accessed 13 March 2021).

Dila, Dilman. 'The Last Storyteller'. In *Afrofuturism*. London: Heady Mix, 2019: 95–103.

Eagleton, Terry. *Literary Theory: An Introduction*. Oxford: Blackwell, 1996.

Grishakova, Marina. 'Poetics of the Return: On the Formation of Literary Canon'. *Senoji Lietuvos literatūra* No. 18, 2004: 23–29.

Guillory, John. 'Canon'. In *Critical Terms for Literary Study*, F. Lentricchia and T. McLaughlin (eds). Chicago, IL: University of Chicago Press, 1990.

—— *Cultural Capital: The Problem of Literary Canon Formation*. Chicago, IL: University of Chicago Press, 1993.

Habila, Helon (ed.). *The Granta Book of the African Short Story*. London: Granta, 2011.

Kakoma, Nyana. 'Dilman Dila on Being an All-Round Storyteller'. *Sooo Many Stories*, 22 September 2014. https://somanystories.ug/blog/authors/2014/09/dilman-dila-round-storyteller (accessed 4 July 2020).

—— 'Acan Innocent Immaculate'. *Sooo Many Stories*, 10 April 2019. https://somanystories.ug/book-authors/acan-innocent-immaculate (accessed 27 September 2020).

LéO Africa Review 'Innocent Immaculate Acan'. 15 July 2019. https://leoafricareview.com/author/acan (accessed 27 September 2020).

Manadue, Christopher Benjamin and Cheer, Diane Karen. 'Human Culture and Science Fiction: A Review of the Literature 1980–2016'. *Sage open* Vol. 7, No. 3, 2017: 1–15.

Milner, Andrew. Review. '*Metamorphoses of Science Fiction: On the Poetics and History of a Literary Genre* by Darko Suvin, Gerry Canavan; *Dystopia, Science Fiction, Post-apocalypse: Classics – New Tendencies – Model Interpretations* by Eckart Voigts, Alessandra Boller'. *Utopian Studies* Vol. 29 No. 3, 2018: 421–429.

Moonsamy, Nedine. 'Aliens and Insecticide: Ecoambiguity in Two Short Stories from Dilman Dila's *A Killing in the Sun*'. *English in Africa* Vol. 46, No. 3, 2019: 75–92.

Murua, James. 'Lillian Akampurira Aujo'. *Writivism* Workshops 2015. www.jamesmurua.com/tag/lilian-akampurira-aujo (accessed 27 September 2020).

Ngũgĩ wa Thiong'o. *Petals of Blood.* London: Heinemann, 1977.
Orwell, George. *1984*. London: Arcturus Publishing, 2013 [1949].
P'Bitek, Okot. *Song of Lawino*. Nairobi, East African Publishing House, 1998 [1972].
—— *White Teeth*. Nairobi: East African Educational Publishers, 1989.
Paton, Alan. *Cry, the Beloved Country*. New York: Charles Scribner's Son, 1948.
Radway, Janice. A. *Reading the Romance: Women, Patriarchy and Popular Literature*. Chapel Hill: University of North Carolina Press, 1991.
Ross, Trevor. *The Making of English Canon: From the Middle Ages to the Late Eighteenth Century*. Montreal: McGill-Queens University Press, 1998.
Ryman, Geoff. 'Lillian Akampurira Aujo'. 100 African Writers of SFF: Part Seven Uganda. *Strange Horizons*, 25 January 2018. http://strangehorizons.com/non-fiction/lillian-akampurira-aujo (accessed 27 September 2020).
Samalar, Sofia. 'Towards a Planetary History of Afrofuturism'. *Research in African Literatures* Vol. 48, No. 4, 2017: 175–191.
Suvin, Darko. *Metamorphoses of Science Fiction*. New Haven, CT: Yale University Press, 1979.
—— 'On the Poetics of the Science Fiction Genre'. *College English* Vol. 34, No. 3, 1972: 372–382
wa Ngũgĩ, Mukoma and Murphy, Laura T. 'Introducing Issue 43: This Hustle is not Your Grandpa's African Lit'. The African Literary Hustle. *New Orleans Review* No. 43, 2017: 1–4.
Whatsonafrica. '"Where Pumpkin Leaves Dwell": by Lillian Akampurira Aujo – Winner of the Inaugural Jalada Prize for Short Fiction, 2015'. https://whatsonafrica.org/where-pumpkin-leaves-dwell-by-lillian-akampurira-aujo (accessed 27 September 2020).
Wright, Timothy Sean. 2019. 'Surviving the African Anthropocene: Dilman Dila's Mutational Aesthetics'. *Research in African Literatures* Vol. 50, No. 2, 2019: 142–157.

Moving the Centre

Positions & Locations of African Speculative Fiction

JAMES ORAO

Ten years ago, approximately fifteen years after the advent of the term 'Afrofuturism', Nnedi Okorafor (author of *Who Fears Death*, *The Book of Phoenix*, *Akata Witch*, *Binti*) asked compatriot and filmmaker Chikere, if Africa was ready for science fiction. In response Chikere points to the difficulty of creating a futuristic imagination, when everyday essentials remain unmet, saying further that 'Africans are bothered about food, roads, electricity, water wars, famine, etc., not spacecrafts and spaceships' (Okorafor 'Is Africa Ready for Science Fiction?'). This brusque assessment of African cultural production results from a single but problematic fundamental discursive script regarding African literature: that the historical trajectories of most African states – of colonialism, engagement with postcolonialism and the contemporary exploitative capitalism, coupled with the socio-political conditions – make for an unfertile ground for speculative fiction. It implies that narratives from and of Africa have other overriding agencies that would fit poorly in a futurist imagination.

Several writers, however, have pointed to the ubiquity of speculative motifs in African storytelling in general; an observation that echoes Ytasha Womack's assertion that she was 'an Afrofuturist before the term existed' (*Afrofuturism: The World of Black Sci-Fi and Fantasy Culture*: 6). That speculative elements in African storytelling are not novelties is further highlighted as a mission statement by *Omenana*, an online literary magazine for speculative fiction whose first issue was published in 2014: 'In our folktales animals talked and the gods walked among men; what is called fantasy today was as real as night and day' ('What We're About'). Not only were speculative elements in storytelling prevalent, but they also formed a core component of African narrative compositions, with natural, supernatural and preternatural motifs, employed as strategies of defamiliarization in the context of social critique. Critics have since recognized these narrative elements as more people now have greater access to Africa and African cultural productions and are better able to hear more stories coming out of the continent. So, while speculative fiction might be new in the canon of 'conventional' African literature – an aspect of cultural production that is dependent on internal and external and sometimes oppositional discourses – their motifs are not new in African narratives. According to this perspective, the definition of speculative fiction in African literature bases not so much on its subject matter but rather focuses on the vast interplay of codes, conventions and, perhaps, archetypes which can be applied to different levels of the corpus (see also Roberts, *The History of Science Fiction*).

For Okorafor, currently one of the most prolific writers in this genre, science fiction remains 'one of the most relevant and potent forms of storytelling'. While discussing the place of speculative fiction in the African literary corpus, she advocates for the genre, commenting on its simultaneous 'distancing and associating effect'; a defamiliarization that makes it 'an excellent vehicle for approaching taboo and socially-relevant yet overdone topics in new ways' (*African Science Fiction is Still Alien*). Lauren Beukes, the author of *Moxyland* and *Zoo City*, concurs with this opinion, adding that 'science fiction allows us the distance to circumvent issue fatigue in our very troubled times' ('The Power of Afrofuturism': 330). One finds this defamiliarization strategy, for example in Okorafor's novel *Who Fears Death*, where the dystopian narrative underpins a critique of the practice of female circumcision as well as the emancipation of the female child, and in *Akata Witch*, where Sunny the female young adult protagonist has to navigate otherizations based on her being a girl, an albino, and a witch in a constellation of motifs and themes that unravels 'vulnerability, victimhood, and ostracism' (Aiyetoro and Olaoye 'Afro–Science Fiction': 235). Speculative fiction, therefore, is not a fringe genre. Following the above exposition, it is just as suitable and capable of dealing with everyday societal issues as realist fiction, but without the didactic undertones that accompany much of realist fiction.

Having testified to African speculative fiction's capability, however, Okorafor decries, first, the dearth of Africans in the 'creative process of global imagining that advances technology through stories' and the lack of 'main characters of African descent, African mythologies, African locales, or address [of] issues endemic to Africa'; and second, the apparent neglect of speculative fiction as a tool of political and social critique in Africa, noting it is a 'literary tool which is practically made to redress political and social issues' (Okorafor, *African Science Fiction is Still Alien*). It is noteworthy that steps are being undertaken to not only address the perceived paucity of this genre in the corpus of African literature, but also to, perhaps, prove to the naysayers that this genre has come into its own. In a recent call for submissions, the Mabati Cornell Kiswahili Prize for African Literature announced 'a special award for 2021 for youth writing science fiction and speculative fiction in Kiswahili, for writers aged between 18 and 35' ('The Mabati Cornell Kiswahili Prize'). This can be seen as a three-pronged approach that, first, seeks to promote the speculative fiction especially among the youth; second, encourages a futuristic imagination of Africa in an African language; and third, endorses the genre as having established itself enough to warrant a prize as well as mainstreaming it. As already argued above in relation to realist fiction, African speculative fiction is in a position to develop new aesthetics that would constitute 'a powerful counter-discourse to the traditionalist images of the African continent' (Rettová 'Sci-fi and Afrofuturism in the Anglophone Novel': 177). This is especially so, because the classic science fiction, having not been made for and by Africans within indigenous African settings, is remote from African experiences and is not capable of adequately imagining African futures.

The conceptual and discursive framework that emerged to engage speculative fiction from Black writers of African origin has been Afrofuturism. However, the corpus classified under this generic term is so diverse as to render Afrofuturism amorphous (see Bould 'Afrofuturism and the Archive, and Delany 'The Mirror of Afrofuturism': 173). Born 'partly out of frustration … at the incredible whiteness of '90s technoculture' (Dery 'Black to the Future: Afrofuturism (3.0)'), the term's 'composite character' allows for 'an elasticity that makes even the oddest interpretations possible': it combines the 'vague Western cultural imaginary of Africa' and a general orientation towards the future (Stompor 'Afrofuturism – A Careful Approach to an Evasive Term') to try to grasp those cultural productions that are both avant-garde and fringe, but that synthesize 'elements of science fiction, historical fiction, speculative fiction, fantasy, Afrocentricity, and magic realism with non-Western beliefs' while re-envisioning the past and speculating about the future rife with cultural critiques (Womack, *Afrofuturism*: 9). Dery, in formulating his original theses on Afrofuturism, posed two questions of the American cultural field: 'why do so few African-Americans write science fiction, a genre whose close encounters with the Other … would seem uniquely suited to the concerns of African-American novelists?' and, 'why are black people, so much a part of the American experience, so conspicuously absent from the stories we tell ourselves, as a society, about technological utopias and posthuman possibility …?' ('Black to the Future: Afrofuturism (3.0)'). These two questions have remained at the core of discourses on the positions and locations of Africans in the speculative fiction genre.

Because of the 'elastic' nature pointed at above and because of its 'composite character', Afrofuturism has been defined variously by different scholars. Dery, who coined it, defined it as 'speculative fiction that treats African-American themes and addresses African-American concerns in the context of twentieth-century technoculture' (180); Womack (9) looks at it as 'a way of imagining possible futures through a black cultural lens'; Dery, in 'Black to the Future: Afrofuturism (3.0)', revises it as 'a way of looking at current events, historical memories, and ... the future through the lens of black speculative fiction'; Lauren Beukes contributing in *Nature* summarises it as 'an artistic, aesthetic and philosophical movement that combines science fiction, magic, traditional beliefs, black history and culture' ('The Power of Afrofuturism': 330); Isiah Lavender III and Lisa Yaszek ('The First Death of Afrofuturism': 2) look at it as a 'reboot of black identity' and a way for Black culture to contest white supremacy; and Kniaź ('Capturing the Future Back in Africa': 53) interprets the term as a cultural and political aesthetic that '(re)imagines the world in which Black individuals are granted the right to speak for themselves, to reflect on their history, and envision Afrocentric futures'. At the core of all these definitions and interpretations lies a desire to establish and locate a specifically African-American niche within the genre in a manner that would be markedly different from Anglo-American-centric pioneering period of speculative fiction.

It is important at this point to first trace a few discourses on speculative fiction of Black origin while reaching out towards the objective of this article, i.e. to present a digest of its positions and locations within the larger discourse on speculative fiction in general. Anderson and Jones, in the introduction to *Afrofuturism 2.0: The Rise of Astro-Blackness*, amended and expanded the scope with the term 'Afrofuturism 2.0', which they define as 'the early twenty-first century technogenesis of Black identity' that reflects on the Black identity in a transdisciplinary approach, and as a 'Pan-African Afrofuturist movement' that takes into consideration the 'regional differences such as, and not limited to, Caribbean Futurism, African Futurism and Black futurism' (x). Afrofuturism 2.0 is characterized by five dimensions:

- metaphysics (engages relations between the ontological and epistemological aspects of knowledge, cosmogony, cosmology);

- aesthetics ('includes anthropomorphic art, music, literature, and performance');
- theoretical and applied science (archaeology, astronomy, architecture, etc.);
- social sciences ('sociology, anthropology, psychology, political science, history'); and
- programmatic spaces ('exhibitions, community organizations, online forums, and specialized salons or labs').

(Anderson and Jones: x)

It is, thus, an organic discursive evolution; the interdisciplinarity moves the original iteration of Afrofuturism into a wider field that allows for discursive incursions from and in different directions but always having a sort of Pan-African approach at its most basic.

Dery's neologism, though contested, was an attempt to tag and, in so doing, help archive those texts that dealt with the 'technoculture' of the African-American. That he posed his question from the premise of a history of marginalization and othering by white American society has, however, been criticized by Hope Wabuke, in her article in *the L.A. Review of Books* titled 'Afrofuturism, Africanfuturism, and the Language of Black Speculative Literature'. Wabuke, who sees Dery's scope of definition of Blackness as too typically white, Western and narrow, accuses him of failing to imagine 'a Blackness independent from any relationship to whiteness, erasing the long history of Blackness that existed before the centuries of violent oppression by whiteness'. While not rejecting the term Afrofuturism, Wanuri Kahiu problematizes it, noting that 'Afrofuturism' is used to tag her work and put her in a specific category beyond generic science fiction ('Africa & Science Fiction'):

> my work is being called Afrofuturist, it is not being called science fiction … to put it in a box where it is understandable to people that it comes from a black person … an African person or a person of African descent

– a labelling she considers as taking away from the inherent aesthetics of her works, and as such negative. To address this perceived anomaly of nomenclature and misleading aesthetic signposting, Wabuke proposes a move towards Okorafor's terms of 'Africanfuturism' and 'Africanjujuism'. It is worth noting at this point that Okorafor has consistently and insistently, especially in her interactions and comments on Twitter, come forth to point out why her writing is not 'Afrofuturism' but 'Africanfuturism', as will be seen shortly.

While Okorafor once regarded her writing as Afrofuturist ('Sci-fi Stories that Imagine a Future Africa') she has since not only revised her position on this classification but gone ahead and coined two terms that she uses to refer to her works: 'Africanfuturism' and 'Africanjujuism'. She notes that the common denominator of all science fiction, regardless of location and position, is their political nature. However, she takes the same stand that has been taken by other writers and intellectuals of African origin and emphasizes the different genealogy of her science fiction writings by proclaiming that 'not all science fiction has the same bloodline, that line being western rooted science fiction … my science fiction has different ancestors, African ones' ('Sci-fi Stories that Imagine a Future Africa'). This assertion effectively locates and positions her work firmly beyond the majority of science fiction that is rooted in a Western Eurocentric tradition. On 2 September 2020, through her Twitter handle @Nnedi, she wrote: 'I write *Africanfuturism*. I do not write Afrofuturism. I write *Africanjujuism*. I do not write Afrofuturism' (@Nnedi; emphasis mine). Indeed, @Nnedi has been vocal, engaging her followers on Twitter on the correct way to 'read' her fiction. The 'different ancestors' locate her speculative fiction away from the Afrofuturist conceptual and aesthetic framework. What is even more noteworthy is her agency in how and where her works are to be located; and that is not in a general imaginary of a Black community that stretches beyond the African continent and into the Americas.

To outline what differentiates Africanfuturism from Afrofuturism, Okorafor penned her thoughts regarding African speculative fiction in her blog, nnedi.blogspot.com, under the title 'Africanfuturism Defined'. She says she coined the term because 'Afrofuturism' seemed to be a catch-all term with 'several definitions and some of the most prominent ones didn't describe what I was doing'; that she was being tagged an Afrofuturist even though those definitions did not apply to her and her works, which led to these 'being read wrongly'; and that she needed to regain how she was being defined. By rejecting the basis of the definitions of Afrofuturism, Okorafor seems to be reiterating compatriot Chimamanda Adichie's sentiments against the 'dangers of a single story'. As quoted above from her tweet, Okorafor considers her works as either belonging to Africanfuturism or Africanjujuism; the move away from 'Afro' to 'African' denoting a move to position and locate her works away from a hyphenated space defined by Blackness solely in relation to whiteness, to a 'purely' African space, both on the continent as well as the diaspora. Therefore, Africanfuturism seeks to emphasize that African speculative fiction defines African realities on its own terms. It is, looked at this way, what Nyamnjoh considers a continuation of the struggle of African intellectual history: a 'struggle of how to reconcile international recognition with local relevance' (Nyamnjoh 'A Preview': 1). Africanfuturism is thus a distinctively African way of perceiving and interpreting African life and thought as to be found in African speculative fiction.

Yaszek's understanding of Afrofuturism helps demarcate the two terms: First, she defines Afrofuturism as a 'black Atlantic intellectual project' resting primarily on questions of slave(ry) history and its attendant 'conditions of homelessness, alienation, and dislocation'; second, Afrofuturism as anchored upon questions of race, while harking back to the questions of 're-interpretations and re-appropriation of an effaced or suppressed past' ('Afrofuturism in American Science Fiction': 58), and last, Afrofuturism as a sort of re-writing project that would seek to re-imagine Blacks in a more just and equitable society . With this demarcation, Afrofuturism is, first and foremost, an African-American storytelling project. The 'Afro' prefix would, of course, include the general 'Black Atlantic' as well, but not continental Africa, which simply remains 'the site of the massive alien abduction of the slave trade' (Samatar 'Toward a Planetary History of Afrofuturism': 176), but still removed from the discourses that birthed Afrofuturism. The diasporic Africa, therefore, remains the location of most of the Afrofuturist texts while Africanfuturist texts would seek to locate their stories within continental Africa. Cultural productions from the African continent, therefore, need to tackle the uniquely continental African issues as well as the continent's relationship to not only the global but also future imagination.

Okorafor's contemplation is, therefore, apt as not only is 'there … currently a lot of discussion as to whether African SF, which is currently flourishing, should be understood in terms of Afrofuturism' (Bould 'Afrofuturism and the Archive': 175), but others have considered Afrofuturism as '"geopolitically inappropriate" as a descriptor of speculative fiction from the African continent' (Samatar: 175). Steingo, too, expresses unease in using the term Afrofuturism to refer to cultural productions from outside of its African-American coinage and prefers the term 'African Afro-futurism' in reference to creative practices and theoretical activity on the African continent ('African Afro-Futurism: Allegories and Speculations': 50). Noting the need to be cognizant of the recent explosion of speculative fiction texts in Africa, while also recognizing the agency of colonial dispossession and expropriation and how this mirrors the transatlantic slavery that produced the same forms of agentic dispossession in the Black Diaspora, he makes quite a problematic conclusion: 'rather than thinking about Afro-futurism in the Americas and Africa as distinct movements, then, it may be useful to establish a more inclusive Black Atlantic narrative' (50). This conclusion is problematic because the shared heritage that links the Black Diaspora to Africa does not, unfortunately, eradicate the disparate experiences resulting from the singularities of the transatlantic slavery and the African colonization on both sides of the divide; issues arising out of these unique experiences must be dealt with at the sites of those atrocities and will, save a few convergent points, require unique discursive and conceptual frameworks.

According to Okorafor, Africanfuturism and Afrofuturism may share a heritage, with the explicit difference 'that Africanfuturism is specifically and more directly rooted in African culture, history, mythology and point-of-view …, and it does not privilege or center the West', and this difference is key ('Africanfuturism Defined'). While the Black Diaspora seeks to maintain and even re-connect with the African continent as the primordial homeland, the disparate history, experiences and destinies between the two render an attempt at comparison difficult. They are not opposites and, however dehumanizing, alienating and disempowering the experiences of slavery and colonization were to the respective groups, the sum of the varying experiences and the proceeding discourses and socio-political peculiarities have ensured differing worldviews.

Okorafor's Africanfuturism equally seeks to place the African in Africa within the prevailing cultural productions based on speculative and futurist motifs; she exemplifies this with the Wakanda reference, as will be seen shortly. Africanfuturism recognizes that the position and the location of Africa and Africans within the global discourse have always been through foreign agency – 'prey to a dominant colonial and colonising epistemological export' (Nyamnjoh: 33) –and a whole continent and nations whose historical orientation has been overshadowed by colonial imperialism and capitalism remained in danger of being further unwritten in the imaginaries of the future as well unless they owned the storytelling – the motifs, the characters, the locations. As Okorafor continues in 'Africanfuturism Defined', in contrast to Afrofuturism, Africanfuturism's 'default is non-western; its default/center is African' and it does not have to extend beyond the continent; but because it is organic, of course, it does. Africanfuturism proposes to look at the African experience – past, present and future –

and to concentrate on an exploration of uniquely African experience within the African continent, as opposed to the historically Black diasporic experience which anchors Afrofuturism. She further argues that Africanfuturism moves the centre to Africa, but is not closed to Africans from outside this centre.

To differentiate between a possible Afrofuturistic approach from an Africanfuturistic approach, Okorafor gives the following example: 'Afrofuturism: Wakanda builds its first outpost in Oakland, CA, USA. Africanfuturism: Wakanda builds its first outpost in a neighboring African country' ('Africanfuturism Defined'). Africanfuturism, in this regard, places the initial points of discursive incursions within the continent itself before branching out into any engagements with the rest of the world; the future imaginary is firmly anchored on the African 'sciences' and 'technologies' which then parley with the other – global or alien – from a position grounded on ontological and epistemological specificities from Africa. With the centre in Africa, Africanfuturism harks back to African mystical elements 'drawn or grown from actual African cultural beliefs/world views' ('Africanfuturism defined').

Okorafor further explicates Africanjujuism as 'a subcategory of fantasy that respectfully acknowledges the seamless blend of true existing African spiritualities and cosmologies with the imaginative'. The rich fabric of cosmologies and spiritual pluralities form the core of Africanjujuism which brings to the fore *pluriversal* – as opposed to universal – and *plurivocal* – as opposed to univocal – sensibilities. Africanjujuism proposes that technology is not alien to Africa and that 'technology is just another form of juju' (Whitted 'To Be African Is to Merge Technology and Magic': 209). In both cases of Africanfuturism and Africanjujuism, the alien and other-worldly are implicitly acknowledged and are woven as part of the storytelling. Universal motifs, storylines about family, growing up, love, chasing your dreams (for example, in Okorafor's *Akata Witch* and *Akata Warrior*) are, in Africanfuturism, firmly rooted in Africa, based in and on specific African cultures, and told from an African perspective respecting the African historical and socio-political constellations.

The term Africanfuturism, of course, will have to navigate a 'newcomer's dilemma' in its endeavour to position and locate African speculative fiction within Africa. What are the points of differentiation? What exactly makes it different from the more than 25-years-old Afrofuturism? And conformity, how does it, in its desired difference, remain within the conventions of speculative fiction? These are weighty questions that considerations of Africanfuturism – as a sub-genre of speculative fiction at par with Afrofuturism and not a subcategory of Afrofuturism, as Kniaź ('Capturing the Future Back in Africa': 54–55) would like to read it – need to consider as they demarcate its space within prevailing discourses, especially in light of the existing Afrofuturist trend. Indeed, this chapter argues that subject matter alone is not enough to differentiate Africanfuturism from the established Afrofuturism; discourse around the term will need to develop and expound on its theoretical and conceptual tenets as well as on its poetics and aesthetics, to circumvent oppositional discourses that might want to view its location as too limiting and, as a result, brand it a tribal or cultural complement to Afrofuturism, such as Kniaź (54–55) seems to imply, regarding Africanfuturism more 'as a particular manifestation of Afrofuturism'. Emerging discourses, especially on the intersecting points of the two terms will have to be reflective enough to consider questions of roots – the rootedness of Afrofuturism in the historical experiences of the African-Americans and Black Diaspora as opposed to the continental African roots of Africanfuturism – and range – here understood as the level of and positions on inclusivity. While Okorafor puts premium on the location and position of telling the African story as the key indicators of belonging to Africanfuturism, Delany considers the question of who is telling the story equally important ('The Mirror of Afrofuturism': 173). Africanfuturism, looked at within these aspects, is then defined further as the speculative and/or futuristic imagining of phenomena, events, ideas, etc. related to Africa from within the African ontological and epistemological frameworks.

For some authors, like Lauren Beukes, Afrofuturism and Africanfuturism are terms that do not hinder the understanding of what they are doing. For Beukes, it is the content creation and codification of the works that is important. Speculative fiction in Africa is important in so far as it is 'preoccupied with the unique social issues … and the creative, alternative and unexpected ways in which people living in them respond' (Beukes 'The Power of Afrofuturism': 330). Even though she uses the term Afrofuturism while referring explicitly to speculative fiction from and about Africa, she is talking about such fiction being in touch with historical and socio-political realities on the African continent. She is, according to the foregoing arguments and prescription, an Africanfuturist. Indeed, her novel *Zoo City* and Okorafor's *Lagoon* seem to employ the same strategy: they both draw from African mythologies, cosmologies, and spiritualities to reinforce the 'perception that what is important in African culture has survived and will survive confrontation with the West' (Gaylard, *After Colonialism*: 92). *Lagoon*, for example, reanimates 'Udide, the narrator, the story weaver, the Great Spider' (228) and her cousin Ijele (291) and Anansi (292), and 'Mami Wata … the goddess of all marine witches' (235) to intrude in the storytelling of the alien landing off the coast of Lagos. The radical break and technological disruption such an alien landing forebode are thus not narrated in isolation of the pre-existing ontology: 'the city and its mythological others meet and merge to form an urban environment of entropy and enchantment' (Hugo 'Looking Forward, Looking Back': 48). *Zoo City*, on the other hand, combines the fantastic '*mashavi*' and '*aposymbiots*' with the traditional spiritual instances of *muti*, *nyangas*, and the *sangoma* (traditional medicine and healing practices and practitioners) to transcend the realist narrative of Johannesburg and to advance 'futuristic visions of dystopian new world orders from local points of view' (Hugo: 47). The juxtaposition of African mythologies, traditional spiritualities and cosmologies as well as worldviews and events in a more modern referential matrix typifies Africanfuturism. For the same reasons that Dery coined the term Afrofuturism to differentiate, then archive speculative cultural production that was uniquely African-American, so does Okorafor's coinage of Africanfuturism approach that same genre from and about the African continent.

The terms Africanfuturism and Africanjujuism are, therefore, exploratory and geared towards conceptualizations of Africa partaking in the crafting and the telling of the African in the future and within the discourses of futurist imaginations. Just as Afrofuturism aims to 'create their own takes on futuristic life through the arts and critical theory' (Womack, *Afrofuturism*: 8), Africanfuturism also seeks to deliberately combine social commentary and entertainment to tell stories in which 'the root of the technology, cultural shifts, sentiments, concerns, characters, way of speaking, needs that drive the story' are 'first and foremost … endemically African' (Okorafor 'Is Africa Ready for Science Fiction?'); the unfamiliar in the African speculative fiction must incorporate the historically, spiritually or cosmologically familiar. Africanfuturism, therefore, seeks to 'place Africa at the vanguard of planetary discourse, producing a new wave of cultural output that signals the continent as a site from which to imagine the emergence of future worlds' (Hugo: 47).

Dery posits a question that is at the core of deliberations regarding the relationships between African-Americans and speculative fiction: 'Can a community whose past has been deliberately rubbed out, and whose energies have subsequently been consumed by the search for legible traces of history, imagine possible futures?' ('Black to the Future: Interviews': 180) To ask the same for Africanfuturism: can a continent whose nations were subjugated, its socialization reengineered, and whose energies have in the last 50 years been consumed by the reconstruction of identities and nationhood, imagine possible futures? And what kind of futures could they imagine, bogged down as they are with the burdens of colonialism and traumatized psyches?

Alastair Reynolds, writing in *Nature* notes that science fiction is the product of agitation. It comes forth when the future is 'less assured, the present less tractable': 'Science fiction is not in the business of reassurance. It is instead dedicated to turbulence, transformation and unpredictability' ('The Bards of Turbulence': 333). Afrofuturism and Africanfuturism both imagine futures where the historical and socio-political realities undergo a subversive re-imagining: they 'appropriate, repurpose, maybe even radically re-imagine the new technologies of interconnection, simulation, and mass reproduction to their own politically subversive, idiosyncratically personal, or historically black ends' (Dery 'Black to the Future: Afrofuturism (3.0)'). For the African-American, writing speculative fiction has always been regaining agency over their own stories to tell about themselves in a bid to shape who they are within their diasporic realities; the same can be said of Africanfuturism in regards to African speculative fiction and its concern with the African condition. It is the locations from which the Africanfuturists begin their quest that is of importance and that makes the difference necessary. Without this differentiation, speculative literature of and about Africa risks suffering under Afrofuturism what the Afrofuturist texts suffered while subsumed under the general category of speculative fiction within the American literary production.

As such, Africanfuturist texts are not too far from the African reality to be explicitly science fiction nor are they too whimsical to be just fantasy; they present real African possibilities in the most extreme light and reflect them in all their absurdity. In fact, I would advocate for the *alien* in African literary productions: a literary imagination that seeks new constellations between aesthetics, history, and space; it is a poetics consisting of complicities and resistances: not necessarily rejecting history in totality, but rather operating with as well as beyond history and refusing to be solely defined by historical agencies, and thus presenting the African condition with decidedly different rules of interaction.

WORKS CITED

Aiyetoro, Mary Bosede and Elizabeth Olubukola Olaoye. 'Afro–Science Fiction: A Study of Nnedi Okorafor's *What Sunny saw in the Flames and Lagoon*'. *Pivot* Vol. 5 No. 1, 2016: 228–246.

Anderson, Reynaldo and Charles E. Jones. 'Introduction: The Rise of Astro-Blackness'. In *Afrofuturism 2.0: The Rise of Astro-Blackness*, Reynaldo Anderson and Charles E. Jones (eds). Lanham, MD: Lexington Books, 2016: vii–xviii.

Beukes, Lauren. 'The Power of Afrofuturism'. Science Fiction: When the Future is Now. *Nature* Vol. 552, December 2017: 330–331.

—— *Zoo City*. Sunnyside: Jacana Media, 2010.

Bould, Mark. 'Afrofuturism and the Archive: Robots of Brixton and Crumbs'. *Science Fiction Film and Television* Vol. 12, No. 2, 2019: 171–193.

Delany, Samuel R. 'The Mirror of Afrofuturism'. *Extrapolation* Vol. 61, Nos 1–2, 2020: 173–184.

Dery, Mark. 'Black to the Future: Afrofuturism (3.0)'.. *Fabrikzeitung.ch*, 1 February 2016. www.fabrikzeitung.ch/black-to-the-future-afrofuturism-3-0 (accessed 24 September 2020).

Dery, Mark. 'Black to the Future: Interviews with Samuel R. Delany, Greg Tate, and Tricia Rose'. In *Flame Wars: The Discourse of Cyberculture*, Mark Dery (ed.). Durham, NC: Duke University Press, 1994: 179–222.

Gaylard, Gerald. *After Colonialism: African Postmodernism and Magical Realism*. Johannesburg: Wits University Press, 2005.

Hugo, Esthie. 'Looking Forward, Looking Back: Animating Magic, Modernity and the African City-future in Nnedi Okorafor's *Lagoon*'. *Social Dynamics* Vol. 43, No. 1, 2017: 46–58.

Kahiu, Wanuri. 'Africa & Science Fiction: Wanuri Kahiu's "Pumzi" Oulimata Gueye', 2009. www.youtube.com/watch?v=SWMtgD9O6PU (accessed 26 September 2020).

Kniaź, Lidia. 'Capturing the Future Back in Africa: Afrofuturist Media Ephemera'. *Extrapolation* Vol. 61, Nos 1–2, 2020: 53–68.

Lavender, Isiah, III and Lisa Yaszek. 'The First Death of Afrofuturism'. *Extrapolation*, Vol. 61, Nos 1–2, 2020: 1–5.

Nature. 'Science Fiction: When the Future is Now'. *Nature* Vol. 552, 2017: 329–333.

Nyamnjoh, Francis B. 'A Preview'. In *Drinking from the Cosmic Gourd: How Amos Tutuola Can Change Our Minds*, Francis B. Nyamnjoh. Mankon, Bamenda: Langaa RPCIG, 2017: 1–32. www.jstor.org/stable/j.ctvh9vw76 (accessed 28 February 2021).

Okorafor, Nnedi. 'Africanfuturism Defined', 2019. http://nnedi.blogspot.com/2019/10/africanfuturism-defined.html?m=1 (accessed 21 September 2020).

—— 'African Science Fiction is Still Alien', 2014. http://nnedi.blogspot.com/2014/01/african-science-fiction-is-still-alien.html (accessed 21 September 2020).

—— *Akata Warrior*. New York: Viking, 2017.

—— *Akata Witch*. New York: Speak, 2011.

—— 'Is Africa Ready for Science Fiction?'. 2009. http://nnedi.blogspot.com/2009/08/is-africa-ready-for-science-fiction.html (accessed 21 September 2020).

—— *Lagoon*. London: Hodder, 2014.

—— 'Sci-fi Stories that Imagine a Future Africa'. *TED* Talks, 2017www.youtube.com/watch?v=Mt0PiXLvYlU (accessed 20 September 2020).

—— *Who Fears Death*. New York: DAW Books, 2014.

Omenana magazine. 'What We're About', nd. http://omenana.com/about (accessed 17 March 2021).

Rettová, Alena. 'Sci-fi and Afrofuturism in the Anglophone Novel: Writing the Future and the Possible in Swahili and in Shona'. *Research in African Literatures* Vol. 48, No. 1, 2017: 158–182.

Reynolds, Alastair. 'The Bards of Turbulence'. Science Fiction: When the Future is Now. *Nature* Vol. 552, December 2017: 332–333.

Roberts, Adam. *The History of Science Fiction*. New York: Palgrave Macmillan, 2006.

Samatar, Sofia. 'Toward a Planetary History of Afrofuturism'. *Research in African Literatures* Vol. 48, No. 4, 2017: 175–191.

Steingo, Gavin. 'African Afro-Futurism: Allegories and Speculations'. *Current Musicology* Nos 99–100, 2017: 45–75.

Stompor, Tomaaz. 'Afrofuturism – A Careful Approach to an Evasive Term'. *Fabrikzeitung.ch*, 1 February 2016. www.fabrikzeitung.ch/afrofuturism-a-careful-approach-to-an-evasive-term (accessed 24 September 2020).

'The Mabati Cornell Kiswahili Prize – The Nyabola Prize for Science Fiction 2021in Kiswahili / Tuzo ya Nyabola Uandishi Wa Hadithi za Kisayansi na Kidhana kwa Lugha ya Kiswahili', 9 March 2021. https://kiswahiliprize.cornell.edu/special-prize-for-2021 (accessed 15 March 2021).

Wabuke, Hope. 'Afrofuturism, Africanfuturism, and the Language of Black Speculative Literature'. L.A. Review of Books, 27 August 2020. https://lareviewofbooks.org/article/afrofuturism-africanfuturism-and-the-language-of-black-speculative-literature/#disqus_thread (accessed 26 September 2020).

Whitted, Qiana. '"To Be African Is to Merge Technology and Magic": An Interview with Nnedi Okorafor'. In *Afrofuturism 2.0: The Rise of Astro-Blackness*, Reynaldo Anderson and Charles E. Jones (eds). Lanham, MD: Lexington Books, 2016: 207–213.

Womack, Ytasha L. *Afrofuturism: The World of Black Sci-Fi and Fantasy Culture*. Chicago, IL: Lawrence Hill Books, 2013.

Yaszek, L. 'Afrofuturism in American Science Fiction'. In *The Cambridge Companion to American Science Fiction*, G. Canavan and E. Link (eds). Cambridge: Cambridge University Press: 58–69.

Featured Article

Reimagining Transracial Intimacy

The Cartography of Decolonial Love in Leila Aboulela's 'Something Old, Something New' & Tomi Adeaga's 'Marriage and Other Impediments'

GABRIEL BÁMGBÓṢÉ

> Today I believe in the possibility of love; that is why I endeavor to trace its imperfections, its perversions.
>
> (Frantz Fanon 2008, 28)

Writings on love in African literary studies have often suffered critical neglect. Most times, such writings are excluded from the economy of 'serious' art, especially when they are short story forms produced by women. In the introduction to *ALT 31*, 'Writing African in the Short Story', Ernest Emenyonu decries the marginalization of the short story form in African literary criticism considering that African writers 'have used the short story to comment on various aspects of life in modern African societies'. Despite the critical neglect, African writers continue to use the short story to signal 'a new sense of direction, moral regeneration, social integration, racial tolerance, and equality of all humans under the law' (6). For African women practising such a significant form that has suffered critical marginalization, contending with the problem of double marginalization is inevitable. Nonetheless, I seek to critically examine how African women writers reimagine love as a decolonial technology for remapping the cartography of the self and the other in the short story form. I examine how they contest the ethno-normative frame of love to offer other possibilities beyond the colonial regulation of social intimacy. I ask: How do African women writers' engagement with the aesthetic act of reworking love foreground other ethical inter-relational orientations that clearly depart from colonial inter-racial intimacy?

To engage this question, I carry out a critical analysis of two short stories in Ama Ata Aidoo's edited *African Love Stories: An Anthology* – Leila Aboulela's 'Something Old, Something New' and Tomi Adeaga's 'Marriage and Other Impediments' ('Something' and 'Marriage' hereafter). I choose these narratives because of their central trope of transracial intimacy. Theoretically positioning these short stories at the intersection of feminist, postcolonial, and decolonial thoughts on love, I consider the decolonial thesis, themes, and conflicts in the stories by analysing how these African women writers imaginatively rework transracial intimacy as a liberating force for transforming the self and the other. Through the trope of transracial intimacy, the Sudanese writer, Aboulela, and the Nigerian writer, Adeaga, radically critique the colonial boundary, the self and the other. I argue that their reimagination of transracial intimacy evokes a double act of routing and re-routing the potentials of *decolonial love* that lays bare the problem of colonial imagination of being (Nelson Maldonado-Torres 'On the Coloniality of Being'). My reading illuminates their appeal to the interconnected reality of the modern world by evoking the power of love as a radical transformative force transcending the limits of gender, class, and racial, cultural, linguistic, and national borders.

In *Black Skin, White Masks*, Frantz Fanon's deep meditation on a mode of 'ethical orientation' he describes as 'a movement of love' powerfully articulates a search for 'authentic love' in a world of colonial violence (28). Fanon clearly understands the power of love in the transformation of the subject even as he struggles with the problem of colonial violence and its dehumanization of black lives. He clearly understands that to overcome the problem of violence, the work of love must supplant colonial aggression. The work of love is a necessary precondition for constituting the new human capable of transforming the traumatic colonial experience in the process of remaking the self. However, the paradox of this ethical mediation on love as a social technology for constituting the new human lies in the fact that Fanon in this critical inquiry fails to accept love in the totality of its imperfections and perversions. Any form of love deemed to be 'imperfect' love becomes damned as *inauthentic*. The damnation of imperfect love in Fanon's meditation is understandable because his project, even as it aspires to envision a new and an ethically transformed human, remains implicitly framed by a kind of masculinist and nationalist imagination that responds, as he has beautifully claimed, to the 'architecture' of his time (5–6). Consequently, it appears that the search for 'authentic love' in Fanon's project offers a *possibility* erasing other *possibilities* of love. That is why Fanon could not help but condemn certain kinds of imperfections and perversions of love in his tracing. However, as Carolyn Ureña has noted in 'Loving from Below', Fanon's investment in the critique of colonial/hegemonic inter-racial liaison opens up the possibilities for rethinking ethical interrelation of being.

African women writers offer an alternative to the colonial hegemonic transracial liaisons presented by African male writers such as Fanon in his critique, Ferdinand Oyono in *House Boy*, Ousmane Sembène in his film, *La Noire de*… or East African Asian writer such as M.G. Vassanji in *Uhuru Street*. By remapping transracial intimacy and relation, the women storytellers disrupt the violent logic of colonial cartography of being, which Fanon describes in terms of the Manichean zones of being and nonbeing that efface the possibilities of ethical contact and the intersubjectivity of being. This colonial cartography of being is the core invention of colonial modernity that continues to structure the global consciousness of the world. The women's reconfiguration of transracial intimacy deconstructs its hegemonic patriarchal, colonial, imperial, and capitalist logic by imaginatively reworking the Fanonian thesis in a way that accepts and recognizes the radical possibilities, imperfections, and perversions of love.

The anthology's blurb positions the stories as evidence of African women's radical imagination, and I should add that this radical imagination has a deep history in their storytelling tradition. As the blurb further notes, the stories in the anthology 'deal with challenging themes representing some of the most complex love stories ever published from Africa'. The decolonial thesis of Aboulela's and Adeaga's creative projects embodies the complexity of love. Their stories demonstrate that there are many forms of love through which ethical relations could be attained, and there are multiple levels of understanding of these forms of love that exceed the limits of the normative frame of colonial imagination of intimacy. This thesis assumes an imaginative scaffold in the working out of the themes and conflicts in the stories. Undeniably, the trace of this thesis can be found in Aidoo's story, drawn from African oral tradition, in the introduction of the anthology. The pervasive love of Achire is contested by her society but she, just like her partner Kwasa, obeys the pull of her heart to the point of excluding herself from her kinship. She defies her father's regal and patriarchal authority to express her love. Like other women's storytelling, which Obioma Nnaemeka, following Trinh Minh-ha, has described 'Mother's Talk' (10–14, 26–32), I argue that Aboulela's and Adeaga's stories, echoing the paradigm of 'Achire's Heart', launch what Chela Sandoval names as a 'neorhetoric of love' in the African imagination (*Methodology of the Oppressed*: 3). Love, Sandoval elaborates, 'is reinvented as a political technology, as a body of knowledges, arts, practices, and procedures for re-forming the self and the world. This affirmative practice and interpretive strategy, this hermeneutics of love easily bypasses the usual order of perception' (4). Indeed, it is in this bypassing of the order of perception that the women writers' affirmation of 'pervasive' love destabilizes the ethnonormative order of the transracial intimacy and such an affirmation promises a decolonial cartography of affinity and relations. Elaborating on Fanon's and Sandoval's theorization of decolonial love, the work of Maldonado-Torres, Ureña, Junot Díaz, Yomaira Figueroa, Joseph Drexler-Dreis, and Sabelo Ndlovu-Gatsheni have articulated how this attitude of love can overwrite the incalculable violence of colonial modernity in order to heal, repair, and rehumanize the world.

Aboulela's 'Something' and Adeaga's 'Marriage' engage in this radical project by approaching the frame of love through the theme of transracial intimacy. In an interview with Paula Moya, Díaz categories transracial intimacy, which he differentiates from 'decolonial intimacy', as one of the masculine logics of freedom questioned by women of colour in their critiques of power and oppression: 'first of all these sisters were pretty clear that redemption was not going to be found in the typical masculine nostrums of nationalism or armed revolution or even that great favourite of a certain class of writerly brother: transracial intimacy' ('The Search for Decolonial Love'). Even though Díaz would not agree that transracial intimacy or what I would rather call transracial love is liberating because it masks the power of oppressive mechanisms, Aboulela's and Adeaga's stories engage in the practice of reinventing transracial love to transform self-other relations.

In 'Something', a story in which we experience the actions and reactions of nameless characters, a Scottish lover travels down from Edinburgh to Khartoum to marry his Sudanese fiancée he met at 'the Sudanese restaurant near the new mosque in Edinburgh' (4). This place, 'the Nile Café', where he first meets the lady is significant. The place constitutes of a Sudanese space elsewhere. First, it is a Sudanese restaurant named after the river Nile, the river that fills his mysterious imagination on his arrival to Khartoum when shown by his fiancée: '"Look," she said, "take off your sun-glasses and look. There's the Nile." And there was the Nile, a blue he had never seen before, a child's blue, a dream's blue' (2). The narrator informs us how proud the Khartoum lady is of this river space even when her lover can only picture savage crocodiles and the fatality of 'blood, death, bones, beneath the surface of the mysterious blue' (3).

Second, this Sudanese restaurant is within the proximity of a new mosque. The mosque draws attention to the lady's religion and identity. In their conversation in Edinburgh, when he tells her that he converts from Catholicism to Islam, '[s]he seemed almost surprised by his answers. She associated Islam with her dark skin, her African blood, her own weakness. She couldn't really understand why anyone like him would want to join the wretched of the world' (7). This narrative represents an entanglement of place and religion as constitutive of identity. It is in this Sudanese place elsewhere that the love affair first begins before its performance on the Khartoum landscape. Lindsey Zanchettin's insight on the articulations of the 'notions of [Muslim] faith and home as a liberating practice' in Aboulela's migration stories illuminates how these notions constitute the African female subject's complex identity ('Articulations of Home & Muslim Identity in the Short Stories of Leila Aboulela': 40). But what critical move is being made by the perpetual location of the white male subject, as a subject falling in love with the Sudanese female subject, in a Sudanese space constituted elsewhere? The logic of the white male subject's location within the space of the other is intended to re-form his received colonial imaginaries about the other so that he could attain decolonial love which creates 'lines of affinity [that] occur through attraction, combination, and relation carved out of and in spite of difference' (Sandoval 168–169).

The conflict of transracial love plays out in the staging of the anxiety of difference. In Aboulela's story, the anxiety of difference and the question of choice and consent find creative expression in the Khartoum lady's brother's reactions to his sister's foreign fiancé: 'He looked irritated. Perhaps by the conflicting desire to get his sister off his hands and his misgivings about her marrying a foreigner' (1). This moment in the narrative stages the scepticism towards the humanity of the other. The choice of words foregrounding his irritation, conflictual impulse to sever the tie between the self and the 'foreign' other registers his attitudinal misgivings and scepticism towards racial difference. We are led to the inner struggle with differential anxiety through the rhetorical question: 'How did he see him now, through those narrow eyes, how did he judge him?' (1–2). There is a perception of the foreign other and a judgment passed on his being based on exteriority. This exterior perception makes relation between the foreign fiancé and the brother so impossible that the brother cannot even accept the foreigner's greetings. Certainly, the foreigner is aware of his fiancée's brother's scepticism; he expresses his unease to her: 'I don't think your brother likes me' (4). Transracial love is unthinkable for the brother who appears to stand in the gap of the voice of his people's family.

In a summative duration, we are taken to the lady's past marriage to a Sudanese man, which the man's family and the lady's brother consent to. The brother, in fact, spends a lot on the bride price and the wedding. The man's family wants a Sudanese girl for their son who falls in love with an English girl because they also cannot accept the transracial love. The marriage collapses and the Sudanese man returns to his English love. The Sudanese lady's Scottish fiancé also experiences failure in his past relation with his same-race girlfriend. The scenes of failure prompt the question: why is it that the same-race love fails when transracial love succeeds in both cases? I would suggest that the failure of same-race love is a critique of the ethno-normative frame of heterosexual intimacy and its nationalist orientation. The narrative vision subverts the logic of racial homogeneity which denies the self the possibilities of opening up to receiving the other. The sustenance of the transracial love pushes us to accept other relations outside the accepted ones. We are opened up to the anti-normative underside of heterosexual intimacy and kinship. This form of love deconstructs the notion of transracial love as a betrayal of one's race or as a revenge on the other.

The sense of love troubles the boundary between the individual and the collective. The narrative's portrayal of the tension between the individual and the collective in relation to the concept of choice in romantic relationship recalls Chielozona Eze's argument on the turn to the individual in African new writing. Eze argues that new African writing 'seek[s] to subvert the obsession with the collective ... [by bringing] awareness that the dignity of the individual is at the center of every community' (115, 117). I would argue, based on how the stories approach the question of the individual and the collective (as we shall see in the other story), that the collective is not necessarily undermined to place the individual at the centre of imaginative consciousness. Rather, the writers imagine a negotiation of the individual and the collective. In Aboulela's story, the sense of the collective disrupts the Western claim of the absolute right of the individual which the foreign fiancé might have brought with him. In every outing and everything the lovers do, '[t]hey were never alone' since there is always the presence of the whole family everywhere they are (10).

Aboulela's story brings the question of loss to the experience of love. On Ghamhouriah Street, the Scottish fiancé is stripped of every possession in a theft incident; he lost his claim to legal identity (passport) and his optical machine used in capturing exterior value (camera). The loss coupled with another loss, the character of the

Bill Cosby look-alike's death, lowers the guard of both groups and engenders an ethical relation between the foreign and the African subjects. I would propose that what each group lost is a symbol of their ego. Just the same way the Scottish man places much value on his passport and camera, his fiancée's family hold Bill Cosby look-alike dear. With the loss of these ego symbolics, we begin to experience a turn of event and real entering into each other's zone of consciousness. The brother comes to accept the foreigner. They walk on the street together and agree on the possibility of the marriage even in the face of loss. The death opens up the space for them to pray together. Afterward, the tension between the brother and the foreigner eases: 'In the car, there was a new ease between them, a kind of bonding because they have prayed together' (14). With this new relation, a new kinship emerges between the different people. The native becomes hospitable to the stranger. Even the stranger began to 'dwell' in the place he has located in his imagination in the primitive time on his arrival. Her place that 'disturbs' him becomes his place of dwelling. At the family's cemetery, '[w]ith the sleeve of his shirt, he wiped the sweat off his brow – he was beginning to act like them – since when did he wipe his face with his shirtsleeves in Edinburgh?' (14). The implication of him acting like them is suggestive of his identifying with them especially when we think of the loss of his legal identity as a Western subject. Although he is still being haunted by the place as seen in his dream of 'a vivid but unclear sadness' complemented by the dream of 'being chased by the man who had ripped his rucksack, stolen his passport and camera' (17), he is able to attain decolonial love and 'through that love', in Díaz's words, 'a decolonial self'. The final tension emerges at the point of marriage when the Imam refuses to contract the bond because the white male subject does not have a Muslim name. The Imam's recourse to 'official certificate' to determine the legality of his religious identity fails as the invocation of voice and community takes on the mode of identity articulation. What Díaz calls 'decolonial intimacy' stimulates the bonding of the different racial communities through the tropes of marriage and loss in transracial intimacy.

In Adeaga's story, told in the first-person voice through the perspective of Tola, the theme of transracial love is represented in the love affair between Tola and her German fiancé, Till. Here we witness another love entanglement between a German male subject and a Nigerian female subject. The 'impediments' of transracial love are presented first through the received narrative of an unnamed female friend recounted to the first-person narrator by her friend, Veronica. Then the narrator takes us to the framed narrative of the self by shifting from the received narrative to self-meditation:

> But I never knew I would find myself in the same situation later. Paradoxically, even my parents vehemently refused to give their consent to the union. They could not bear the thought of their daughter getting married to a German guy. Looking back at the whole incident now, I do not fail to see the disparities born not only out of differences in language but also in cultures and mentalities. This is my story. (22)

This narrative shift at once launches the story as a work of memory and its shock, and the conflict of difference engendered by transracial love. The narrative critiques of the anxiety of difference which arises from the conflict of language, culture, and the perception of the other. This critique orients the narrative frame as a kind of testimony as the marked ellipsis signals to the reader that the story represents a slice of the narrator's personal life.

Tola, a twenty-seven-year-old daughter of upper-class Nigerian parents sent to Germany for study where she later settles and falls in love with a German, Tillmann Sutherer. The affair was treading towards marriage. However, the difficulty of Tola relating her intention of getting married to a German is expressed in two rhetorical questions in free direct discourse: 'How am I going to tell my dad that the guy I've chosen is not a Nigerian, but a German? What am I going to do, I ask myself?' (22) These questions reveal the narrator's awareness of her people's apathetic attitude towards transracial love. Tola begins to strategize how to handle the 'unusual' situation in her mind before travelling to Nigeria. She soon realizes that she is caught up within the Yorùbá tradition, where the choice of a marriage partner is not the sole judgment of the lover. Even though she has a choice to make, the choice also depends on familial consent: 'I realized that a lot depended on my family giving me their consent, because no matter how long I may have lived away from home, I still cannot do away with my traditions. It has been the most important part of my education' (25). The narrator's statement conveys the presupposition that being far away from home is not in any way a disconnection from home. The significance the narrator places on tradition profoundly articulates a sense of feeling at home with oneself and one's tradition. Thus, the challenge for her is not how to reject tradition but how to pierce through the fabric of tradition in order to make it accept this 'unusual' love.

'Marriage' portrays the paradox of 'constrained choice' in the issues of love and marriage, which we slightly experience in 'Something'. This paradox hinges on the negotiation of individual choice and collective consent. Aissata Sidikou and Marame Gueye in their work on nuptial songs in the Sahel regions of Africa establish that marriage is the bonding of not only individual lovers but also of families and communities. The mode of bonding or kinship that marriage forges beyond the individual, 'whether traditional or contemporary', makes it 'a complex institution in Africa' (Sidikou, *Recreating Words, Reshaping Worlds*: 32). Hence, the choice of the individual in the constitution of this complex institution becomes regulated by kinship ties. If choice means the freedom to choose, this freedom is not only the freedom of the individual, but it is also a form of freedom overdetermined by the collective. Gueye explains that even though there is 'the possibility of choice on the part of the prospective bride … the choice is not completely individualistic and private' (73). With the choice of the individual in the narrative comes the constraint of the family. The lady in Aboulela's story and Tola in Adeaga's make their choices of their lovers who are white male subjects. Nevertheless, the choice of the individual confronts the consent of the family and community. The Scottish lover in 'Something' has to journey to Khartoum to stand the test of the community choice despite the fact that '[h]er country disturbs him' (1), as the story opens, because it takes him through the journey of memory to the primitive time that bolsters the landscape with the imaginaries of bareness, austerity, and stasis, imaginaries in conformity to the colonial Western imagination of Africa as a barbaric space. As J.M. Coetzee writes, 'Africa could never, in the European imagination, be the home of the earthly paradise because Africa was not a new world' (2).

In 'Marriage', this imagination of Africa as an uncivilized space is at work in the conceptions of Till's parents (the middle-class Europeans, Margit and Klaus) of Tolu as an African. Till's mother's remark invokes the trope of jungle reeking with poverty and incivility: 'How can someone whose family lives in a jungle somewhere in Africa think she's good enough for my son? … I will not have you bring shame into our family by tainting our blood with this woman' (24). And when Till implores them to know his '*Negerin*' (in his father's phonetic rupturing of 'Nigerian') fiancée before drawing conclusions about her, his mother replies:

> We don't have to do so because Africa is a poor and uncivilized continent where the inhabitants depend on aid from hard-working German taxpayers. Once you get married, you'll have all those hungry looking villagers asking you and your family for aid. (24–25)

For Ursula, his paternal grandmother, the African (female) subject cannot fall in love; she can only take advantage of the liaison to claim papers (European citizenship) through witchcraft. The imagination of the primitive African cannot be separated from her primitive practices of witchcraft.

Again, the anxiety of difference plays out in the colonial imagination of racial difference. On the side of Tola's family, her sister, Titi, expresses shock in her sudden lapse into Nigerian Pidgin when Tola confides in her that her fiancé is a German. Her mother appeals to the discourse of public face and shame in Yorùbá to express what she perceives as the impossibility of such transracial relations. Tola's father's statement signals the question of individual and collective choice: 'Although we know you're of age, we are still responsible for you. Remember that, in our culture, when you marry someone, you also marry the whole family' (28). Her father's reference to marriage as a bonding of communities affirms Sidikou's and Gueye's argument on kinship and marriage. Tola has to defend her choice by appealing to the intercultural reality of the modern world and the power of love as a force beyond the control of gender, class, racial, cultural, linguistic, and national barriers. Indeed, it takes the intervention of a family meeting for Tola's choice to be accepted, an intervention presided over by the wisdom of the most senior member of the extended family, Papa. The role of Papa in the story reaffirms the role of the elders and the power of 'seniority' in the mediation of marital choice (Gueye '*WoyyiCeet*: Senegalese Women's Oral Discourses on Marriage and Womanhood': 72).

Nonetheless, the anxiety of difference dissipates through ethical contact. The distance inhibiting the real understanding of each other has to be bridged. Till's parents are shocked when they realize that 'they were unwanted' by Tola's family before the turn of event (32). As a result, they grow the curiosity to know the other, a project which they initially reject. The wedding takes place in Lagos. Till and his parents travel to Lagos to meet with Tola's family who meet all their needs. Till's family realizes their stereotype of Africa as an uncivilized jungle reeking with poverty, dependent on the West for aid is false. Through the contact and the bridging of distance, decolonial intimacy is forged between the different people. I should note that the narrative act of journeying to the 'the native land' of the other undertaken by the white subjects/strangers, not as conquering subjects but relational subjects, in Aboulela's and Adeaga's stories reorients the violent history of colonial contact and intimacy. This reorientation of contact and intimacy takes the narrative project beyond the realm of mere romantic love to decolonial love. Their creative vision of ethical contact critiques the aggressive contact of colonial penetration of the native land. The movement to the space of the other is a strategic lowering of white supremacy to the ethical level ground of hospitable intimacy and intersubjective understanding.

In conclusion, African women writers in their short stories remap the cartography of love with the aim of pushing the boundary of racial difference. These stories offer us alternative frames through which we can understand and recognize transracial love. Aboulela and Adeaga engage the question of transracial love like Fanon and Díaz but they ask us not to limit this mode of love to racial betrayal and revenge. In the context of these women's writing, this mode of love is not a product of 'alienation', the inability to accept oneself as a result of a sense of inferiority, 'a neurotic orientation' that is 'pathological' (Fanon: 43). Rather, it is an expression of choice and agency, as constrained or otherwise as they might be, in the interconnected world. This mode of love fosters ethical relations across different races, relations that engender a thorough understanding of differences through the ethical orientation Maldonado-Torres designates as 'decolonial attitude' ('Cesaire's Gift and Decolonial Turn': 127). This love is not only the integration of individuals but also of their different racial communities in a 'moral order predicated on decolonial love and its ethics of all humans living together harmoniously' (Ndlovu-Gatsheni 'Geopolitics of Power and Knowledge in the COVID-19 Pandemic': 19). The happy endings of both stories, which in a way recuperate the formulaic endings of fairy tales, launch us into the space of decolonial utopia and transformation. Recognizing and accepting love in its imperfections and perversions is significant to unlocking the true possibilities of freedom, an 'ethical orientation' that human consciousness needs to transcend the tendency of a 'movement of aggression' (Fanon: 28).

WORKS CITED

Aboulela, Leila. 'Something Old, Something New'. In *African Love Stories: An Anthology*, Ama Ata Aidoo (ed.). Ayebia Clarke: Banbury, 2006: 1–21.

Adeaga, Tomi. 'Marriage and Other Impediments'. In *African Love Stories: An Anthology*, Ama Ata Aidoo (ed.). Ayebia Clarke: Banbury, 2006: 22–33.

Aidoo, Ama Ata. 'Introduction'. In *African Love Stories: An Anthology*, Ama Ata Aidoo (ed.). Ayebia Clarke: Banbury, 2006: vii-xiv.

Coetzee, J.M. *White Writing: On Culture of Letters in South Africa*. Braamfontein: Pentz Publishers, 2007.

Drexler-Dreis, Joseph. *Decolonial Love: Salvation in Colonial Modernity*. New York: Fordham University Press, 2019.

Emenyonu, Ernest N. 'Editorial: "Once Upon a Time Begins a Story…"'. *African Literature Today 31*. Woodbridge: James Currey, 2013: 1–7.

Eze, Chielozana. *Postcolonial Imagination and Moral Representations in African Literature and Culture*. Lanham, MD: Lexington Books, 2011.

Fanon, Frantz. *Black Skin, White Masks*, trans. Charles Lam Markmann. London: Pluto Press, 2008 [1952].

Figueroa, Yomaira. 'Reparation as Transformation: Radical Literary (Re) Imaginings of Futurities Through Decolonial Love'. *Decolonization: Indigeneity, Education & Society* Vol. 4, No. 1, 2015: 41–58.

Gueye, Marame. '*WoyyiCeet*: Senegalese Women's Oral Discourses on Marriage and Womanhood'. *Researches in African Literatures* Vol. 41, No. 4, 2010: 65–86.

Maldonado-Torres, Nelson. 'Cesaire's Gift and Decolonial Turn'. *Radical Philosophy Review* Vol. 9, No. 2, 2006: 111–138.

—— 'On Coloniality of Being: Contributions to the Development of a Concept'. *Cultural Studies*, Vol. 21, No. 2–3, 2007: 240–270.

Minh-ha, Trinh. 'Mother's Talk'. In *The Politics of (M)Othering: Womanhood, Identity, and Resistance in African Literature*, Obioma Nnaemeka (ed.). London and New York: Routledge, 1997: 26–32.

Moya, Paula M.L. 'The Search for Decolonial Love: An Interview with Junot Díaz'. *Boston Review*, 26 June 2012. http://bostonreview.net/books-ideas/paula-ml-moya-decolonial-love-interview-junot-d%C3%ADaz (accessed 18 March 2020).

Ndlovu-Gatsheni, Sabelo. 'Geopolitics of Power and Knowledge in the COVID-19 Pandemic: Decolonial Reflections on a Global Crisis'. *Journal of Developing Societies* 2020: 1–24.

Nnaemeka, Obioma (ed.). *The Politics of (M)Othering: Womanhood, Identity, and Resistance in African Literature*. London and New York: Routledge, 1997.

Oyono, Ferdinand. *House Boy*, trans. John Reed. London: Heinemann, 1966.

Sandoval, Chela. *Methodology of the Oppressed*. Minneapolis: University of Minnesota Press, 2000.

Sembène, Ousmane (dir.). *La Noire de…* New Yorker Video, 1966.

Sidikou, Aissata. *Recreating Words, Reshaping Worlds: The Verbal Art of Women from Niger, Mali, and Senegal*. Trenton, NJ: Africa World Press, 2001.

Ureña, Carolyn. 'Loving from Below: Of (De)Colonial Love and Other Demons'. *Hypatia* Vol. 32, No. 1, 2017: 86–102.

Vassanji, M.G. *Uhuru Street*. London: Heinemann, 1992.

Zanchettin, Lindsey. 'Articulations of Home & Muslim Identity in the Short Stories of Leila Aboulela'. *African Literature Today 31*, Woodbridge: James Currey, 2013: 40–51.

Interviews

Journeys into the Subterranean

Mystical Realism in Chigozie Obioma's Novels Interview with Chigozie Obioma

LOUISA UCHUM EGBUNIKE

Born in Akure, Nigeria to Igbo parents, Chigozie Obioma's writing draws on the rich cultural heritage of the Yorùbá town he grew up in, his Igbo ancestral home and the English language he writes in. His novels *The Fishermen* (2015) and *An Orchestra of Minorities* (2019), both shortlisted for the Man Booker Prize, immerse the reader in a recognizable, yet mystical world. It is this combination of realistic portraits of places and historical episodes, layered with a palpable spiritual dimension, which has led Obioma to describe his work as 'mystical realism'. Obioma introduces his reader to particular ways of seeing the world, rooted in Igbo and Yorùbá belief systems and worldviews. Rather than narrate details of these cultures, Obioma immerses his reader into worlds in which the spiritual realm and the land of the living, coexist on the same plane. For Obioma, his writing seeks to present the world as the communities he writes about experience it. His acclaimed debut novel, *The Fishermen*, set during the turbulent years of General Sani Abacha's rule in Nigeria, follows the story of four brothers who receive a devastating prophecy, that the eldest brother, Ikenna, will be killed by one of his younger brothers. The mythic and the real merge as the brothers' relationship breaks down and their lives are forever altered. *An Orchestra of Minorities* is narrated by the ancient *chi*[1] of the protagonist, Chinonso, who must make a case to Chukwu, the Supreme Being, following a series of tragic events involving Chinonso. Chinonso risks everything he has for Ndali, the woman he loves, and is left with the trauma of losing it all – including Ndali.

This interview opens up conversations on the generic category, 'mystical realism', created by Chigozie Obioma in an effort to remedy misreading of his work. It considers the paradoxes that are at play in Nigeria, the intersections of the 'real' and the 'imagined' and Obioma's current and future writing projects.

LE: One of the first questions I actually wanted to ask you emerges out of a conversation we have had before when I remember you discussing your work and calling it 'mystical realism'. I wanted to ask you for a definition of 'mystical realism'.

CO: Yeah, yes. Well, so I've given some thought to the sub-genre, let me use that term, of, the kind of fiction that I've been trying to write or have written so far. Especially in long-form fiction. In my short fiction, I generally write more realist, modernist stuff. But I've just been thinking a lot about the ways in which people have miscategorized my work, especially *The Fishermen* and then *An Orchestra of Minorities*. Many have called the books magical realism. I started revisiting some of the earlier influences, of course, Achebe's work to some extent, but also some of the Greek myths and some of the oldest stories in the Igbo tradition, *Omenrk* and some of the older folkloric tales, and then a lot of Nollywood. So what I found is that, modern fiction, Nigerian fiction has actually been centring more of a Western vision or Western mode of thinking, so this is a serious discordance between the kind of stories that we tell in Nigeria, and the kind of fiction that we put out. So there's a question of intellectual hierarchies that have unconsciously formed. So if you have, even if it's just a university education in the West, there's an internal switching, that people don't seem to be conscious of, but it does happen. So you will see a novel like … just because she is my friend let me use Chika Unigwe's *On Black Sisters' Street*, for example. I can never, you know, go wrong with Chika (laughs) … so, so you see a novel like that which is mostly a realist novel, is set in Nigeria, but it's like something that an American or a British person would write. But if we were to tell that story in Nollywood there would be a kind of mystical dimension to it. There would be a reason why these young women decided to do this thing. It could be as a result of some kind of curse or some missteps in the spiritual plane. So there's a factoring in of the sociological thinking and cosmological thinking of the Igbo people, which at its very core, makes the potent, in my opinion, argument that there really isn't any difference between the world of the physical reality and the plane of the ethereal or the hyperreal. They blend together. This is why we can speak of ancestors being around us, this is why we can talk about the concept of the *chi* intervening in our life. This is why we believe in prayer. This is why we do juju. This is why we give blessings some pronouncements, because we believe that in everyday life, there is a direct influence of the supernatural. So that is accounted for in our everyday life, but it's missing in our fiction. You see what I'm saying? So, I am trying to bridge that gap, so to speak. So, so then I have a form that is at least epistemologically Western, the novel, but what we are doing is that we are appropriating a form that is Western, I mean, they began to write the novel. So, we are using a framework that is theirs. So, because of that framework, they are the ones who have the authority of naming things. They are the ones who looked at South American fiction, and said, 'well, this is fantastical, this is magical realism. These are not real, but they write it as if it's real' – you see the thing. So, you can understand why they would say that. So, then I have to look at what they have already named, and on the list: speculative fiction, magical realism, none of them seem to suit this mode of seeing the world, you know, that you see in Nigerian film. Take the work of Ernest Obi for instance, all the films that he has directed. You see that remarkably, almost at every juncture of the story, there are invisible spirits that are hanging around; they just are there. You know, people are eating dinner, and there are spirits roaming around. This is how we see the world. But one thing I noticed also is this: there is a paradox in all of these things, and that paradox is what saves or prevents, to better say it, the Igbo worldview from lapsing into magical realism. You can have all of these spiritual interventions, but there's still free will. There's still the world of the realist, which means that you can leave and do stuff and these things will not interfere with you. I mean they are there, but you can overcome them. So that is what has produced my refined view that what Nollywood produces, and what I have done in my two novels, really is a kind of a realism, but it's a mystical one, so that's the term. To define it, I would say that it's a kind of genre of fiction in which there is a direct or an equivalent emphasis on both the realism and the metaphysical. So both of them are on an equal plane.

LE: That's really interesting. Some of the things you touched on are things I definitely want to tease out. In both of your novels, you are looking at the spiritual, supernatural realm, not as something separate from the realm of humans but as occupying the same space; the ancestors are not remote but very much present. Your works look at both the lives of humans and the presence of the spiritual and, the intersection and interaction between the spiritual realm and the realm of people. Can you speak a bit more about that relationship and the extent to which, in your own works, a tension emerges between the inherent autonomy of humans and the spiritual forces at play which impact their lives. You have spoken on the fact that the difference between the kind of work you're writing and say magical realism is that your human characters have a degree of autonomy, but still the spiritual very much impacts on their everyday lives.

CO: So, if I'm getting your question correctly, that's like the paradox of life in Nigeria and I think that has to be accounted for in our fiction. For instance, I like to use again familiar people, for example to illustrate my point, so, my dad would be a good example of this. This is a very Western oriented individual; there is a skeletal figure of him in *The Fishermen*. So, this is a guy who has a PhD equivalent in accountancy. So he's a very meticulous guy. He would come in here and he would note that something is askew almost immediately. He has extraordinary interest in Westernization and he is one of those who completely reject our ancestral religion and embrace Christianity. So, so he's a good example, but even so, with that complete orientation that points Westward, he believes in the potency of reincarnation. He would argue to the core that I am an avatar of his uncle, and he would give you convincing reasons, and I believe that they are convincing. So if I, of course, put on my Western rationalist hat, I am going to just say no, this is fabulist thinking but no. If you look at Americans, they believe in Halloween. They have magical thinking, too, but because of the economic and political and of course, military power that they have, because of how they dominate most things, their viewpoint is what becomes supreme in the world. All right, so this is to say that he makes to at least in my view, convincing arguments to why we should believe in reincarnation, which is something that is radically opposed to the Western viewpoint. He also believes in some of the gods in my village and their potency. He believes, for instance, that there are certain things that if you do, maybe you go on fish in a stream called *iyi ocha* in his hometown, white river, where the fishes and the wildlife in there are prohibited, you cannot fish them; he believes that yes, if you fish them, something bad will happen to you. There are consequences, spiritual consequences to them. So, that paradox of this hybrid of a person who is both a Western person but at the same time is not, is reflective of a great percentage, probably at least 85 per cent, of not just Nigerians, but I think most Africans. So, I think that you should account for that in your fiction, which means that there will be questions.

You can see in *The Fishermen* for instance, this supernatural world stands almost removed, from the lives of this boy. So Abulu speaks this word that is supposed to be a kind of a curse, right? So if the curse is not something that you see, what is the mechanism for a spell to come to pass? People in Akure, in that novel, just believe that well, somehow what this guy says, happens. If you say that somebody is going to have an accident, what orchestrates it into fulfilment is anyone's guess. But it does happen. So what they see is the result which is physical; the manifestation of the supernatural. Even though they believe in the supernatural, they are much more focused in the material consequence of it. So that material consequence is what I'm most interested in, in my fiction.

LE: It's interesting listening to you talk about your father. On the one hand, Westernized and on the other hand, deeply rooted in Igbo spiritual systems. This paradox, as you put it, and hybridity, speaks to what you're producing as you said, writing in the novel form, which is essentially a Western form, but thinking about ways in which your writing can still encapsulate the kind of mystical worlds you're depicting. So I think that inherent paradox is present, interestingly, in the production of your fiction and in the lives of your characters. And, I just wondered if you could speak a bit about yourself, your experiences growing up and what draws you to writing these kinds of stories in these kinds of ways. What influences did you have that led you on the path to writing 'mystical realism'?

CO: That's a good question. So, my interest at the beginning of my creative life was mostly to tell interesting stories, obviously. I just was more interested in stories that are not based on animals, that is as a child, I just saw more of symbolism in those stories. So as a child, I didn't like to read about tortoise (laughs), you know, or the lion and all that stuff. I was much more interested in the human beings and their stories, and I saw something genuinely wondrous in being able to create a universe that is similar to what we have, but that didn't already exist. So you create a character, you create a world that we can recognize, but at the same time is completely a fictive structure that is original and exclusive to your mind. So I was very interested in that, which I'm sure would necessarily have led me to being a realist fiction writer, but I think the point at which I had a kind of break was when all of the readings I was doing, and all of the stories I was hearing, started to blend together. So I was born in Akure by Igbo parents. So I was speaking Igbo,

Yorùbá and then English and I was reading in all three languages. So, there was a point which at which all of these things intersected. I was reading Amos Tutuola's *The Palm-Wine Drinkard* while reading Achebe's *Arrow of God*, Elechi Amadi's *The Concubine* at the same time I was reading Fagunwa's *Ògbójú Ọdẹ nínú Igbó Irúnmọlẹ̀*, and then I was reading *Macbeth* as well, as you know, *The Iliad*, *The Odyssey* and all of these other Western stories, *As You Like It*. So, at a point, just like Achebe said, I think I found myself at a crossroad. I recognize that man, I am not a particular thing, I am various. Colonialism and postcolonialism and the structure of multi-nationhood in Nigeria has produced a kind of a gallimaufry of things in me. So the English and Western influences are in my prose, in my language. I love the English language. In my fiction, I try as much as possible to push the boundaries of language there. But then the Igbo and Yorùbá worldview became of utmost interest to me. And the stories I wanted to tell, I wanted them to account for those belief systems.

LE: Do you see yourself continuing to write 'mystical realism'? Can you speak about your current book project?

CO: I'm writing a novel on the Biafra war right now. I'm trying to cover a gap, because there's been wartime novels, fantastic novels, but they mostly are interested in the situation during the war, you know, the suffering and all of that. But there's almost no war novel, by which I mean, the fighting because every other situation is a consequence of that. If the federal troops never broke the gridlock at Enugu axis, and the folks at Port Harcourt held their ground, there wouldn't have been the suffering that we experienced, you know the wartime situation in Biafra. In fact, if the Midwest operation had succeeded ... So, the, the situation in the trenches is the most important aspect of war. So, not only am I trying to cover that gap, but I've almost finished the first draft and I stopped at some point and I started to think, look, this just feels inconclusive to me. You know, even though, yes, I have all the battles written, well explored, visually and all that. But the Igbo say something; they say that the story of any war is not only told by those who survive, but also by the dead, because the war's major product, the success of the war depends on how many people you kill. You see, so, if you are able to finish an entire battalion of the enemy forces, that is success. So it's major product, the article of production, is death really, and then only the living tell the story. So if that is the major principle of how war stories should be told, why are we telling these stories by just somebody who survives at the end of the story then accounts for what has happened? No. So how do you have the dead tell the story? I don't want that saying to just be like a crazy saying, you know, something that people just said. No, they believed it. If you woke my grandfather up, he would tell you, 'Yes'. Back in those days, we used to hear the spirits, you know, those people who have been killed. The valiant warriors, they told us this was how I died. So why do we, you know, have to dismiss it. So this is what happens to me – I get to a point where I always think, okay, the lens by which I'm telling the story is not enough. So I see insufficiency in it, and I think that's what gets me to look for different ways of telling the story.

LE: I was just thinking about a couple of things in, in your discussion of your current book project on the Nigeria-Biafra war. One of the things was, if the success of war centres on how many people you kill on the other side, then the depth of the loss is also linked to the numbers dead, and of course, with the Nigeria-Biafra war, it wasn't just death in battle, it was death through starvation as well. So I just wonder, the idea of giving voice to those that have died, particularly in a conflict where many people were never buried properly, which in turn created a kind of disruption to the spiritual order – how does that influence the realm of the living? I wonder if we can, go back to the conversation about the intersecting space of the spiritual and the physical, if you like, if we're hearing the voices of those who have departed, particularly those who haven't had their proper burial rites. What does that tension look like between these spirits who haven't been laid to rest, and those who are left with that heavy sense of loss. Can you speak to that dynamic?

CO: Yeah, no, it's true. I mean, you understand the root of that dilemma. Let me put it that way, because, in the Igbo consciousness, even now, even pastors, ardent Christians who have completely rejected ọdịnanị, will tell you that yes, is if you do not bury someone properly, their spirit does not go to rest, and I think it is true, because, we are not animals. So why must it be that somebody has to be put in the ground? You know, again, it's because we feel like a dead person needs to rest, there has to be some kind of resting place for the person. So, to be able to realize the voices of those people, I have obviously interviewed some people, I've done some research, but I have let my imagination do most of the work. And, I think that the liberty that it gives me is okay. I don't like the idea of having like a cast of characters – I mean it's possible to develop all of them equitably, but there is there's always strength in concentrating your power, your thoughts on one character and then using that character to understand the world as a whole. So, if you have one character who is attached to a specific brigade or battalion or whatever, they are going to be fighting on specific fronts. So, if they are at Nsukka in September '67, they are not at the same time in Port Harcourt where there's, another fighting going on, they are not in Ogoja or any of the other northern frontiers. So how do you account then, for extraordinary monumental and historical events like the Asaba massacre? It would not sound plausible if you should take your character through all of these things. So by actually having some of these dead people come and tell their story, you can have somebody who was at Asaba who was murdered in that gruesome way. I want it to be again just like *An Orchestra of Minorities* – I want it to be a cosmos where you actually are immersed in the war, nothing is missing. You're reading this novel, it's as if you are living all across Biafra between '67 to '70, until the very end of the war. So that is what I'm trying to do. So I am able to reach the places that I couldn't reach through their voices. There are variations of that proverb you know, the one that Achebe used to say that also appears as an epigraph in my book, you know, the story of the hunt glorifies only the hunter, you know that sticks out to me. And, I think, there are ways to do it, so I paused and I'm doing it now. Bad things were done to women who were just civilians; there are some who fought. I have encountered one who actually did fight, but those who didn't, some of them were raped and killed by Nigerian forces, but even some Biafran forces. So even if you have fifty characters, you can't tell all their stories, but actually having some of them being summoned from the grave, you can have a kind of a fuller picture. So, in that sense, it makes for a richer discourse.

LE: Listening to you talk about your new book and thinking about it in relation to your other novels, you said you had to imagine so much of it; could you talk me through the process of building these worlds in your novels. Yes, you're drawing on a particular set of knowledge and beliefs that exist within the Igbo world or within the Yorùbá world. But as a creative writer, you're also having to fill in many gaps and put your own take on it. So could you talk through how you build the worlds of your novel, both in terms of creative processes but also the research that you're doing?

CO: So, one thing I have to say, as a preamble to answering that question is, the interesting thing about this mystical realism, which you're trying to put into a branch of speculative fiction, is that you are using the material that is available to you in the world. *The Fisherman*, the Akure that is there is the Akure you would have found. Even the previous governor of the state has acknowledged that there is a very clear portrait of this place. But then, you have the mysteries, the undercurrents that form a different layer. So it's the same with the Umuahia of *An Orchestra of Minorities*. Chinonso goes about his everyday job, but his chi, is there doing his thing. That is the one distinction. I'm not creating a kind of a fantastical world like *Lord of the Rings*. Because Westerners have dispensed with a lot of these beliefs: they see them as fabulist thinking now, so they don't have that hybrid worldview, that we still have, that hybridization that is a result of the clash between these two cultures, the African and the European. So, in their mind, those things can only exist in an external world that is completely divorced from the reality. So, you don't really see short stories that are fantastical or speculative in the Western tradition, because you need at least 200 pages to build that world. So these books, if you notice science fiction, they are like tomes, *Harry Potter*, because you need to construct a world that is completely unique in your mind, even *Game of Thrones*. So this is to say that even the Biafra situation that I'm creating, I'm not creating anything new. So that's one interesting thing about that. So what you now do is you explore what feels like the subterranean to everyday people. These things exist, we just neglect to see them. It's just like when a priest of ọdịnanị will come in here, and will say, 'ok I can see things that you cannot see'. So as a writer, I am just trying to open my third eye, and see. People who fought in my village, for instance, tell the story of how the river, would sometimes turn into bees. So the Nigerian forces are coming and then it turns to bees, and then it stings them or it drowns them. You see their bodies. Even a Nigerian Hausa soldier did give this account, you know, that it happened.

LE: Finally, you mentioned before that you're working on a novel I think set in fourteenth-century Igboland. Are you able to say anything about that? Because I'm just thinking about where you're going with the different mystical engagements and with these different subjects. I'm wondering what that might look like. Are you able to say anything on that?

CO: I have the plot in my head and I actually have published a poem out of it. I always give a foretaste of what I'm working on by producing something that is related. And you know, it would look like *Arrow of God* you know, and stuff like that. Only that it is earlier. If you watch a lot of Nollywood you probably would have a sense of the kind of thing that I mean, not that I think those films are great – I think that there are lapses. Even though many of them are trying to tell those historical stories and I really appreciate that, but there are also gaps there that they are not accounting for. So this is the first novel that would have a female protagonist; I've been accused of being a gatekeeper for masculinity (laughs) you know, and what not. It's really about this woman whose husband lives to fight a war, they fight with the Arochukwu people. So this is before the confederacy was formed, but the earliest iteration of it. It's not a wartime thing. He leaves. He moves away and disappears, while the rest come back. And she's trying to navigate life with a young child in this society. So I think it will be more like *Arrow of God* and this Igbo trilogy thing. So it's like really a kind of Dickensian exploration of everyday life in an Igbo village or town, so, it would be insanely historical and it will look at the minutiae of life, like every day. What she's thinking, what she is doing and all that, so the purpose is to create a sense of being immersed in this time and in this place, in this culture, in these people, you know, to animate them so that everything comes alive. I mean, *Things Fall Apart* does something similar but it's narrow in a way, because even though it touches all the aspects of life, you know, marriage happens; we see different festivals, seasons of the year, farming and all of these quotidian everyday things, but it zips through this. This one will observe. It's like Knut Hamsun's, *Hunger*, the Norwegian classic, and I think that will be the kind of thing so. It might not be the kind of stuff that people want to read. But you know, I'm going to do it.

NOTE

1 Within Igbo cosmology it is believed that each person has a corresponding and conscious spirit entity.

Translating Magic Realism in Ngũgĩ wa Thiong'o's The Perfect Nine

Interview with Ngũgĩ wa Thiong'o

KADIJA SESAY

The Perfect Nine: The Epic of Gĩkũyũ and Mũmbi is the Gĩkũyũ origin story, which Kenyan writer, Ngũgĩ wa Thiong'o decided to retell in the form of a novel-in-verse – the first time that he has written in this hybrid-genre. Although Ngũgĩ is known as an established playwright, novelist and non-fiction writer with thirty-seven titles that include *Weep Not Child* (1964), *The Black Hermit* (1974) and *Decolonising the Mind: the Politics of Language in African Literature* (1986), he is nevertheless, willing to try new forms to break his own boundaries. In doing so, Ngũgĩ uses the re-telling of the story to create his own form in telling, that has been described as Homeric.

The ten clans of the Gĩkũyũ people are created from the ten daughters of Gĩkũyũ and Mũmbi, who become the Matriarchs of the House of Mũmbi. (The tenth daughter has a physical disability and cannot take part in the challenges with her sisters, yet she has a significant role in the telling of this tale.) Their daughters, Wanjirũ, Wambũi, Wanjikũ, Wangũi, Waceera, Nyambura aka Mwĩthaga, Wairimũ, Wangarĩ also known as Waithiegeni, Waithĩra also known as Wangeci, and Wamũyũ, each have their special characteristics, outlined as they are introduced. As a myth, it is classic speculative fiction and therefore, in the narrative, like other myths and fairytales, the literary style is magic realism. Ngũgĩ's originality is rendered through its complex weaving of layers of forms, styles and stanzas that is a signature of novels-in-verse yet, in doing so, the various layers also reflect the specifics of the story itself, such as using stanzas of ten lines (as there are ten daughters), or multiples of the same in a combination of tercets and quintets, which develop its own rhythmic narrative, first in Gĩkũyũ. As he made the decision to then translate it into English, Ngũgĩ' essentially had to write the book a second time since considering that the syllable counts alone would be different in English, and to ensure that the English translation also developed its own rhythm and, as Ngũgĩ' describes it, its own musicality, only the structure was maintained. This is part of the complexity of writing verse-novels that parallels the complexity of layers in the time, language and location of speculative fiction.

On reading *The Perfect Nine* comes the realization that hints of the Gĩkũyũ creation story are apparent in several of his earlier works of fiction, whether through the name and characteristics of one of the daughters or in aspects of the story presented in a contemporary form. Providing this complete story of the origin of his people is symptomatic of his bringing the bibliography of his novels full circle, demonstrating how poetry, fiction and prose can be produced seamlessly in one piece of work.

The most important facet of Ngũgĩ's telling of the story of *The Perfect Nine* is the philosophy and wisdom that it imparts for the twenty-first century, that which is indubitable and part of the essence of myth.

KS: When did you first hear the story of *The Perfect Nine*? Is this the original title?

NWT: *The Perfect Nine* is the English translation of the Gĩkũyũ original: *Kenda Mũiyũru*. It is based on the Gĩkũyũ myth of origins, traceable back to the original man, Gĩkũyũ, and original Woman, Mũmbi. I first heard it as a child. Each Gĩkũyũ child belongs to one of ten clans. And each clan takes its name from one of the ten daughters of Gĩkũyũ and Mũmbi. So, I grew with the myth. References to it appear in my novels, especially *The River Between* and *Weep not Child*.

KS: What was it about *The Perfect Nine* that you wanted to write and recreate in your own style?

NWT: I was sitting on a raised ground overlooking the Pacific Ocean, when the myth visited me. It reminded me that Gĩkũyũ and Mũmbi had ten daughters and NO male child. So the daughters had no brothers to rely upon. They had to do every task. Make things, including weapons, for their own survival. Theirs was a creative union of mind, hands, and soul.

KS: Myths are associated with ancestral pasts. Do you think that they need to be changed in their telling or format to make them more relevant for the present and future or do you think as myth, the world moves around myth rather than the myth changing to suit new paradigms?

NWT: In my own case, I had to develop certain things inherent in the myth. For instance, in the original myth, when the daughters reach marriageable age, Gĩkũyũ goes up to the top of Mount Kenya and communes with Ngai/God about suitors. The following day, he woke up to find ten handsome young men in the yard, outside the door. In the myth, it says, 'God provided'. But in Gĩkũyũ religious culture, we attribute every success to Ngai/God. So I asked myself, where did the suitors really come from? I saw them as coming from all parts of the continent, ninety-nine in all. So how did they end up the final ten suitors? The tests the suitors undergo are at the centre of the epic.

KS:Do you remember the first Gĩkũyũ myth that you were told as a child and who told it to you?

NWT: This [*The Perfect Nine*] very myth of origins. The teller could have been any of my four mothers, whom I write about in *Dreams in a Time of War.*

KS: Have you come across any parts of this story in other storytelling cultures?

NWT: Maybe in Homer, *The Iliad*. The hundreds who come to woo Helen.

KS: What is special about the creation myth?

NWT: The centrality of women in our being.

KS: What do you believe is the strongest lesson that we can take from *The Perfect Nine* for our lives today?

NWT: Patriarchy tends to lower the place of women in history. *The Perfect Nine* says: 'No, women are the real makers of human history'. Remember, every man and woman in the world was carried in the womb of a woman for nine months. And then the nurturing, that comes with mother's milk. *The Perfect Nine* restores the Woman to her rightful place as the maker of the human race.

NGŨGĨ WA THIONG'O AND HIS WRITING IN THIS GENRE

KS: How do you describe or define your own writing that sits outside realist literature?

NWT: I like the term magic realism. Myths of origins are really in the tradition of magic realism. One of my earliest readings was the Gĩkũyũ translation of the Old Testament. The stories are pure magic. Remember the story of Jonah in the belly of a fish? Or Daniel sharing a pit with lions? The stories thrilled me. Even the story of Noah and the floods.

KS: Do you think that writers are moving away from the realist modes of writing?

NWT: I think that magic realism has impacted African writing. I am a good example of this. My first four novels, *The River Between*, *Weep Not Child*, *A Grain of Wheat* and *Petals of Blood*, are of the realist tradition. People and places could be like any other people and places in real life. But then, in 1977, I was arrested and put in a maximum security prison. I wrote my first novel ever in Gĩkũyũ, on prison toilet paper. It is called *Devil on the Cross*. This is the beginning of my magic realist phase. The inner spirit of the characters may correspond to real life, but the main action and characters do not correspond with real life. Like the main plot: characters from all over the world gathered for a competition of robbers, in a devils cave, to compete to see the best robbers in the world. The devil is a character in the novel. These exaggerations continue in the novel, *Matigari*, to become the dominant mode of narration in the novel: *Wizard of the Crow*, and in the epic, *The Perfect Nine*, in which the characters are confronted by an invisible ogre. They can't tell its size, colour, or even its location.

LINGUISTIC TRANSLATION OF *THE PERFECT NINE*

KS: is translation more difficult when working with a story rooted in tradition?

NWT: Translation from one language to another is always challenging. This is because every language has its own musicality, one unique to itself. This musicality is what makes every language special. It is very difficult, almost impossible, to reproduce the musicality of one language into the musicality of another. Languages are like musical instruments. The piano, the guitar, and the violin can play the same melody but the textures of their musicality are still different. And that is why there is no language which is more of a language than any other.

It is in writing this epic that I truly discovered the musicality of the Gĩkũyũ language.

There is also musicality in the English version, the musicality of English, but each musicality is shaped by the sounds of each language.

KS: Are there figures and concepts that exist within the Gĩkũyũ worldview which have no simple equivalent?

NWT: Yes, absolutely. The word for God in Gĩkũyũ is Ngai. But the word God does not quite capture the notion of Ngai. It comes from the notion of Giver/the universal sharer.

KS: Can you be more expansive and re-write your own interpretations if it cannot be a direct linguistic translation?

NWT: Sometimes, you have to. In my novel, *Wizard of the Crow*, I could not render the word, '*Korwo*', into English with one word. In Gĩkũyũ, the word '*Korwo*' has two meanings: 'If' and 'They will catch you'.

TRANSLATION OF FORM – *THE PERFECT NINE* AND SIMILAR STORIES

KS: What decisions did you have to make when breaking up the narrative into different stanzas and syllable counts for the English translation to work in a similar fashion?

NWT: It is really an epic, or call it a narrative in verse. I had to find a rhythm and verse structure that best suited the story. Very little rhyme. The daughters were ten, but often referred to, as Perfect Nine. I broke the lines and even syllables, where I could, into tens or fives or nines and threes, any multiples of nine or ten.

The main characters are the nine or ten daughters of Gĩkũyũ and Mũmbi. So I use ten and nine, or multiples of the same, for example five, or three or six, for the lines. And even for the syllables in a line. As for the English version, I used the same pattern, but it does not always break as well as it does in the Gĩkũyũ version.

KS: Talk about the rhythm. How different is the rhythm of storytelling?

NWT: This differs from writer to writer. I define a good storyteller as the one who is best able to create anxiety of expectation, and fulfil it.

KS: Writing a myth in a mix of prose and poetry, was it like it is in music, flowing from one verse or form to another?

NWT: I would rather call it verse. I note that some reviewers have compared the verse form in *The Perfect Nine* to Homeric verse.

KS: Is there a process that takes place when moving from an oral tale to that of remixing it and reproducing it in novel form (albeit in verse)?

NWT: No, no there is no formula. You just try your best to make the story flow. In that sense it is more like orature, where the narrator might break into a song, or even into drama.

KS: Are you more storyteller than theorist? What brings the two together for you or are they totally separate? How do you move between the 'realist' and 'speculative' in your body of work.

NWT: I move between the two and also drama. In my lectures, even on theory, I often move from the personal to the theoretical. See for instance, *Decolonizing the Mind.* It is theory and story. Story is theory in time.

KS: What is real, in myth?

NWT: All myths contain a grain of truth.

KS: Can storytelling save us?

NWT: Story is a work of imagination. But story nourishes the imagination. And without imagination, there is no human civilization.

The Steady Rise of African Speculative Fiction

Interview with Chiagozie Fred Nwonwu

KUFRE USANGA

Chiagozie Fred Nwonwu, who writes fiction under the pen name 'Mazi Nwonwu', is the co-founder and managing editor of *Omenana*, a web-based magazine that is home to African speculative/sci-fi/magical realism/fantasy fiction. In this interview, Mazi Nwonwu talks about African SF: then, now, and the future.

KU: Thank you, Mazi Nwonwu, for this interview. In the world of African speculative fiction, *Omenana* occupies a prime position as the resource for SF from all over Africa with over fifteen issues. Thank you to you and Chinelo Onwualu for contributing to the genre's growth and spread since 2014. What inspired you and Chinelo Onwualu to create the online literary magazine Omenana? Take us through the steps that led to creating the inaugural edition.

MN: As a fan of speculative fiction, I continually felt under-represented in the field. I wanted to write distinctly African speculative stories, but getting anyone interested enough to publish them was a struggle. Publishers here used to add a caveat that they won't accept science fiction and fantasy, and even some literary awards specifically said: 'don't submit if your work is speculative'. There was a need for a medium that will cater to the genre, and I was hoping someone would do it. I remember speaking to the late Kenyan writer Binyavanga Wainaina (who was a big speculative fiction fan) about the need to have a magazine that will give writers like myself a platform to write and publish the type of stories we like to read. I recall him asking me: 'why don't you do it?'. During the 2012 edition of the Farafina Creative Writing Workshop (where I met Binya), organized by Chimamanda Adichie, twelve of us produced a manuscript for an anthology we later named 'It wasn't exactly Love'. Chinelo Onwualu was the editor we hired to work on the collection. I wrote a futuristic story, and because she is also a speculative fiction fan and writer, we connected. During one of our conversations, I told her I am planning to set up a magazine that will focus on the genre. She told me she is very interested and will like to be a part of it when I go live. Aside from me having the initial idea, Chinelo went along for the ride. We talked about names, focus, looks, and all that. Even our strong interest in having female voices came from her. Our first edition was published so long ago that it is a struggle to recollect what went down. I know we reached out to some writers and also had a call out. Being players in the genre literature scene in the continent also meant we knew speculative fiction writers. We reached out to many people and had a very healthy inbox to pull stories from. I recall I reached out to a former colleague, David Motutu, and he agreed to illustrate the edition for a pittance. My long-time friend Emeka Okeiyi designed the downloadable version's layout and has since remained a part of the magazine.

KU: Running *Omenana* these past six years, what is your observation on the public's acceptance of African speculative fiction? What has changed in the reception of speculative fiction in Africa over the years, if it has changed at all?

MN: I believe that the perception of speculative fiction in Africa has changed, in that writers of the genre in the continent are more respected now than they used to be. African writers of the speculative are also more visible now, and when you mention Africa and speculative fiction, you don't stop with the work of the cerebral Nnedi Okorafor but will – if you want to do justice – talk about so many other writers with links to the continent doing great things and getting recognition. The African public has always been accepting of speculative fiction by Africans. The thing is that they perhaps don't term it speculative fiction. Our Nollywood movies are dense with elements that are nothing but speculative that tries to answer questions like: 'what if people can be hexed by witch doctors?' 'what if preachers can heal people, make people rich, save people from the clutches of the "devil?"', etc. So, speculative fiction isn't strange to the Nigerian public or reader; they just don't give it that tag. Also, you know much of what we call speculative is realism for the African? So, on Lagos streets, people don't laugh when a man screams that his manhood has disappeared; they take it seriously and start looking for a culprit or solution.

KU: What sub-genres in African SF do you encounter more often, and is there a cultural pattern to why that might be?

MN: Spiritualism, old gods, retold folktales, people wielding otherworldly powers in the cities, etc. are the themes we see more often. Sadly, not so much science fiction as we would have preferred. I don't know if there is a cultural pattern, because people write what they are familiar with. The other-world stories, magic, and spiritualism (I know these terms are deficient, but I've never been able to gift them better expressions) are part of our traditional storytelling form. It is what Nollywood splashes on our screen – even if they don't term them speculative fiction. It is the themes we are most familiar with, and it is the one that the gatekeepers wouldn't readily dismiss as speculative fiction. Then there is the question of the future. You'd be surprised to find that there isn't as much dystopia as you'd think coming from African science fiction writers. Some have said that this is because of how incredibly optimistic Africans are. In summary, you can say that the themes explored here (Africa) are as varied as those in the West, even if the focus differs (no vampires and werewolves lurking in the street corners of Lagos).

KU: In an editorial you wrote on *Omenana* on 1 October 2015, you lamented the crass treatment of Africa in the world of speculative fiction. You admonished that 'Africans need to tell their own stories as this is the best way to own the narrative of Africa and to capture the changing face of the continent'. Has the genre achieved this? Five years later, what is your reassessment of the SF we tell?

MN: We are telling our own stories. The West looked to Wakanda when they woke up to the need to include Africa in their cinematic universe party. Still, most African writers I've met aren't keen on the tale of an invented people being the go-to narrative for Africa. We are writing our own stories. We are harnessing our folktales, legends, or urban legends and making something that is wholly ours. I am thrilled with the stories we are getting. I think this is one of the aspects that *Omenana* can take pride in. Reading the stories on the site, writers tell me, give them ideas of what works, and betters their stories. The genre is taking off, with more writers embracing it, and their abilities to weave beautiful tales are astounding. We are fortunate to be a home for many of these writers.

KU: And speaking of home, the title of the magazine exhilarates me. I understand it is an Igbo word that loosely translates to the English word, culture. Could you unpack this for us?

MN: *Omenana* is an Igbo word. It is culture; it is divinity, and the sum of a people's way of life. I remember Chinelo and I struggling with a name for the magazine. We had agreed to start the magazine, agreed on the format, but what was missing was a name. I recall I was initially struggling to remember the Hausa name for storytelling, but since both Chinelo and I are Igbo, I started thinking of the Igbo word. *Aktru* didn't ring very accurate in my ears, and I said '*Omenana* is a way of life, the SF published here will cover all that'. We agreed on *Omenana*, and it serves us very well. *Omenana* is a magazine that we felt writers of speculative fiction in Africa needed, mostly as many African writers of SF complain about the absence of platforms that cater to their craft. As a writer of speculative fiction, I struggled with finding a place to publish my work. As such, there was a need. *Omenana* filled that need. Now, there are so many other outlets, and we are pleased with where we are now. Hopefully, it will get better.

KU: Are there continuities or discontinuities in the genre's written form when juxtaposed with the oral form of Indigenous storytelling and tales? Are contemporary writers of African SF blurring boundaries and holding and or creating new spaces?

MN: I think the oral form of storytelling will remain with us, so there are continuities. However, since we increasingly are no longer sitting in our mothers' rooms listening to oral tales and narratives passed down from generation to generation, we are passing these stories on in our way. *Omenana*, and other magazines like it, are the new African mothers' room where stories take form. Yes, the borders are being blurred as oral forms increasingly find their way into written works. There are thousands of spaces where this storytelling is taking place in various forms. To discover them, one just has to know the keywords to search.

KU: How has it been managing the demands of running the magazine and payment to writers/contributors, graphic designers, illustrators, and editors to further their craft and keep SF alive and thriving in Africa?

MN: It has been crazy, but I slowed down seriously since 2017 when I changed jobs. The demands of my 9–5 make it very hard to give *Omenana* as much time as it wants. I am thinking of doing it exclusively because that's the one thing that will make me happy. I can't: I've got a family to care for. As such, *Omenana* suffers. We used to pay steadily (from my coffers, lol), but I've found I couldn't sustain that because of family commitments. The plan is to see if *Omenana* can pay for itself, hopefully. However, *Omenana* will remain, striving to give space to the writers who want a space to showcase their talent.

KU: Of course, the need to hold space for African SF cannot be overemphasized. Tell us about the support you have received thus far from organizations and individuals.

MN: Omenana received a grant from the Science Fiction and Fantasy Writers of America, which funded one edition of the magazine. Zimbabwean writer Tendai Huchu gave us his Nommo Award winnings. Wole Talabi also sent us some money, and Goethe Institut gave us funding for our special flash fiction edition. We are grateful for these contributions and how they eased the burden on our pockets when they came. Readers and lovers of SF can support by donating to magazines like *Omenana* to help us pay artists, editors, and contributors, and run the magazine. That's the only way the magazine can remain free for everyone, which we like to see. We want the magazine to get to everyone who can access the internet.

KU: In your estimation, what does the future hold for African SF and vice versa? What role do you see *Omenana* playing in this?

MN: Speculative fiction is an African art form. It was in our past. It is in our present, and it will be in our future. That's assured. *Omenana* will be there, as one of the vehicles carrying the present into the future.

KU: Thank you very much, Mazi Nwonwu, for having this conversation with me.

Literary Supplement

Poison for the Dogs

ESHITIKA L. LUTOMIA

Namukhula tsinuni
Namukhula tsimbande[1]

I graciously sang, hopeful that Namukhula could discern the pleading in my voice and calm down. I even rubbed her neck in the process but she still kicked, knocking down the small bucket I had placed under her udder in readiness for milking. I don't know what was wrong with her but I suspected that Kuka, my little brother, had mixed something in her dairy meal.

"Kuka, Kuka, come here" I shouted but, as I expected, there was no response. Kuka could be somewhere, in the house or the granary or even in the kennel tending to his ever-riotous dogs. I knew he was hearing me but he was clever enough to know that most of the time I called him, it was not for sweets and chocolates. A few minutes later, mama, who was in the smoke-filled outside kitchen preparing evening tea, came out. She was still wearing the same soil-laden *kanga* she had used in the farm. The cassavas and potatoes she had uprooted were well packed in a sisal sack, waiting for me to take them to the market.

"Nafula *baane!* Girls are not supposed to shout like that. What man will marry a woman with a throat as hoarse and loud as a whistle? Kuka has taken his dogs for a walk" mama said, the last statement coming out smoothly as if taking the dogs out was some kind of a revered ritual. As mama wiped her eyes to clear off the tears triggered by the incessant smoke, I realized that everything seemed awry that day: Namukhula was kicking hard every time I touched her udder; the firewood mama was using was not well dried, creating a tent of smoke that engulfed the grass-thatched kitchen roof, rising towards the sky like some jinn escaping from a bottle; and now mama was drawing back to her usual marriage narrative to pin me down.

"What has he done to Namukhula? I think I know what he has done," I asked and answered at the same time like Sherlock Holmes would do when inquiring something from Dr. Watson and realizing that the latter had no clue. Kuka was more stubborn than any of the notorious boys I knew in our village Eshianda-Mushimuli. Mama always said that, that was how most boys behaved when they reached class seven. Mama did not want to believe that Kuka could be smoking the same bhang he used to give to his dogs. He always said that bhang was good for the dogs as it made them more ferocious to scare away thieves who might attempt to steal our cow or goats or even the green maize and potatoes and cassavas in our *shamba*. But to tell the truth, that weed only made those filthy animals madder. Since the death of my father 6 years ago, mama had always been too soft on Kuka. However, she was constantly strict on me, arguing that girls were like eggs and they needed to be guided well. But the way she handled that little brat could even make one wonder who between us was 18 and who was 12.

"I know he has mixed Namukhula's dairy meal with cannabis leaves" I seethed, looking at my mother who had now controlled the smoke in her eyes. Kuka had once told me that our neighbor's cow was prolific because it was always fed on cannabis leaves. I could swear that that was what he could have done when I told him to give Namukhula dairy meal as I prepared milking equipment.

"I swear if I catch him, he will know the kind of fire Nafula breathes," I fumed after sensing no reaction from mama, despite having explained to her the bizarre thing that Kuka had done.

"Don't worry Nafula," she said complacently. "Kuka cannot do such a thing. I know what is disturbing Namukhula," she said as she headed towards the calf pen, bringing out Nana, who started mooing uncontrollably at the sight of her mother. Nana ran towards Namukhula and started suckling vigorously, shooting his mouth forcefully and tactfully towards the udder. After a few suckles, Mama pulled Nana away to prevent him from finishing the milk. Then Namukhula became cool and normal like all days and I bent down to wash her udder once again as I prepared to milk.

Namukhula tsinuni
Namukhula tsimbande

I carried the milk inside the house to pack it into one-liter bottles before taking it to Eshianda-Mushimuli shopping centre. One had to be careful while pouring the milk in the bottles for the last thing I ever wanted was a quarrel with mama. As I filled the last bottle, one of Kuka's dogs crept behind my back and barked loudly, scaring me to the point of spilling the milk. It made me feel like one would feel when walking on the road absentmindedly and then a car hoots suddenly and loudly. Before I recovered from the shock, the bully-dog had already taken off, wagging its tail maliciously, in some kind of primitive victory. I quickly mopped the spilled milk before mama could see it and packed all the bottles in a green paper bag. Mama helped me load the sack half full of potatoes and cassavas on my head. With the sack well-balanced on the head and the milk bag in my left hand, the right hand was left free to swing to propel me forward. I smiled when I remembered my Physics teacher saying that women needed to thank Sir Isaac Newton for discovering the force of gravity, for it enabled them carry loads on their heads without struggle. I wondered if Sir Newton could even balance an empty bucket on his head, let alone that full of water as we girls always did when coming from the stream. As I left for the market, I heard muffled laughs and saw Kuka, his left hand covering his mouth, his eyes turning away from me like a child who had just been caught peeping at his mother's nakedness. "Those dogs", I silently said, the load on my head restraining me from running towards that little brat. Perhaps putting my hands on him could have melted the lump of anger that had solidly accumulated inside my throat.

When I reached Eshianda-Mushimuli shopping centre, I sold all the milk as fast as I had brought it. People liked our milk because, unlike other sellers, we never used to adulterate it as most people did by "adding water." Some neighbors even melted Blue Band margarine and added it into the milk to make it thicker to allow more water. My history teacher blamed it on capitalism. While studying imperialism and colonization in Africa, our teacher said that business dishonesty and crookedness had been brought about by the white people who only cared about money. Before that, Africans placed humanity and hospitality first in all their business transactions. The potatoes and cassavas also went quickly because they were fresh, still smelling soil. After counting all the money and making sure that it was all there, I bought the commodities mama had sent me. I then went to the shop to buy Red Rat, a poison used for killing rodents. However, I was not going to use it on rats. Kuka's dogs had become a menace even to the people they were deemed to protect. In one way or another, the dog madness had to end. I wrapped the remaining money tight at the hanging end of my blouse and headed towards home. Mama would keep it inside her tin where she kept her other moneys and when the schools open, she would buy Kuka new books and a new pair of school uniform. And then that money would finish in a click like she had not been saving at all. Kuka was lucky that the primary school education was free. I was also lucky because my high school fee had been paid from the Constituency Development Fund (CDF) kitty. Mama always prayed everyday that when Kuka join high school, he should also be sponsored through the CDF. But with the increased corruption and bribery claims, one could no longer count on the government for sponsorship.

Walking home from the market with a paper bag bulging from the weighty beef, *sukuma wiki*, kerosene, bones for the dogs and other light commodities made me feel happy. The thought of the deliciously boiled beef spiced with sour milk and the traditional soda powder made from dried banana peels teased my salivary glands. However, what I longed for most was watching Kuka's devils ravish the poisoned bones to the last. With them gone for good, I would no longer worry about watching my back all the time, fearing that those weed-smoking monsters might sneak on me in my bedroom, bathroom, or even in the middle of the shamba weeding maize.

I found mama chopping firewood and wondered why Kuka was not assisting her.

"Where is Kuka?" I asked as I took the paper bag to the kitchen.

"He is in the kennel spraying his dogs. He says that they have been infested by ticks," she replied, her conviction spiking me for I started to believe that mama loved Kuka and his dogs more than me. She had just forgotten that it was the same grubby cannibals that had raided the kitchen two days ago and gulped down the chicken that was roasting on the *meko*.

"Does he think that his dogs are more important than other things? He has done nothing since morning, except caressing those filthy animals," I said as I emerged from the kitchen after unpacking most of the things, bones and the poison sachet the only remaining commodities in the bag.

"Alright. Let us see if tomorrow he shall also spend the whole day with his dog friends."

"Nafula *baane*! Don't be so bitter with him," mama said as she balanced a log of wood to chop it. "He is just a small boy and this is just a stage."

"*Sawa* mama," I said as I entered the house, smiling in the young darkness. I went into my room and locked the door, placed the paper bag on my study table and switched on the light. I quickly retrieved the Red Rat from the piece of newspaper it had been tightly wrapped in. I was not interested in reading the instructions; I just wanted to spread all the powder on the bones and then watch those filthy dogs eat to death. I think that was the most joyous thing I was doing that year, besides sitting for my Form Four national exams. Before opening the sachet, I started taking the bones out of the bag, spreading them on the piece of newspaper that had contained the poison. As I lined the first piece, my snooping eyes kept on darting from column to column on the piece of newspaper and finally, they caught something that evoked my curiosity. The biggest problem with my eyes was that, they never wanted to let anything readable go by, even if it was not pertinent. Most of the columns advertised properties for sale, petty jobs, fertility and vitality doctors among others. What caught my eye however was the column titled "Looking for Form Four Leavers to Work in Classic Hotels in the Middle East."

Since my childhood, I had always dreamed of being a cateress in a large hotel. Before he died, my father had once taken me to Lakeside Hotel in Kisumu city where I spent most of my time admiring the men and women in white coats, their elegance and softness as they graciously delivered trays of edibles to their customers. I cannot forget how I kept on gazing at the large television screens and the gigantic fans spreading wind all over the expansive room. But it was the high school trip to Hilton Hotel in Nairobi that drew my dream closer to my heart. I had never seen anything like that before in my life, a building towering high in the sky with shiny glasses and marbles all over with well-designed rooms. After completing school, I tried to look for employment in the only *kibanda* hotel in our shopping center, without success. I needed to raise my college fee for it was my dream to attend Utalii College in Nairobi, the best institution that offered catering courses. Yet, here was an advert, calling for interested people to work in classic hotels outside the country. I was amazed by everything: working overseas, being posted at a large hotel, a lucrative salary of Ksh 20,000, and the provision of transport costs to the destination. There was a number at the bottom of the advert which interested people could call. I quickly took the newspaper and rushed to the outside kitchen where mama was preparing supper. I was welcomed by a sweet aroma of the steaming meat and for a moment, I nearly forgot what had brought me. I politely asked her to dish me a piece from the golden-brown meat on the *sufuria*.

"Girls are not supposed to behave like this Nafula. How do you think your future husband will react when he is told that his future wife can gluttonously finish all the food when it is still cooking?" she asked as she dished out a fatty piece from the *sufuria*, biting it first before giving me the rest. This time around, I did not care about the marriage narrative.

"Mmmh mama," I said as I chewed slowly to elongate the deliciousness of the succulent meat. "You are the best cook."

"Every woman is supposed to be a good cook. They say delicious food is a way to a man's heart," she said as she swallowed the bit of meat in her mouth. "I see it is ready. Let me add some sour milk and then leave it to boil for five minutes."

"Let it boil fast, my salivary glands are flooding. And before I forget mama, I need to borrow your phone."

"Nafula *baane*! You know I am not giving you. How many times have I told you to stop…"?

"It's not what you think mama. Look, here is a job advert that is paying Ksh 20,000. And you know where? Outside the country. It means I will have to board a plane," I said excitedly, but if I had expected the mention of an airplane would excite mama, I failed miserably. She did not respond immediately. She was either absorbing what I had just told her or she wanted me to complete what I had started. I assumed it was the latter.

"From this advert, all the costs from here to the destination will be paid for. All I have to do is call that number."

"And what makes you think it is true? That, these people you are talking about will employ you?" mama asked as she focused her eyes on the newspaper, even though I was sure that the words on the paper were just decorations to her, like the clay patterns on the walls of our house.

"It is true mama. Look at Anguba. She went to Saudi Arabia and built her parents a big brick house. She said that when she comes back again, she will buy her elder brother a car to start taxi business.

"And has she ever come back? It has been over five years. Her mother says that she no longer communicates. They are afraid that something…something not good might have happened."

"Mama *baane*! Stop thinking like that. Maybe she does not want to be disturbed. Her friend once told me that Anguba was tired with the monthly demands placed by her parents, brothers, uncles, and even some former primary school teachers."

"Let us hope so. But have you not heard stories of families that have lost their loved ones in the Gulf? Let's not go far. Just a few meters from here, Mzee Juma's family. Their daughter came back from the Middle East and died two days later. Postmortem results showed that she suffered much internal bleeding, which could have been caused by daily beatings."

"Those are isolated cases mama. Maybe when such girls reach there, they want to transfer their filthy behaviors from Eshianda-Mushimuli to the Middle East. Who knows? Please mama. Just help me with your phone." I said triumphantly as my mother now retrieved her phone from the black pouch hanging around her neck. I made the call without delay and a lady received at the other end. I told her about the advert and how I was interested in the vacancies described.

"How old are you?" she asked

"I am 18 years. I have just completed secondary school," I replied.

"That is good. Can you travel to Nairobi tomorrow?"

Before I could respond, the call was disconnected and then I realized that mama's *okoa jahazi* credit had finished. Miss Njagi, as she had introduced herself, called back immediately.

"Don't worry about the transport costs Nafula. I will send you the bus fare to Nairobi."

"*Sawa*," I said, my rising voice betraying the overflowing joy.

The conversation did not last long. Miss Njagi told me to call her immediately I reached Nairobi so that she could come pick me. Although mama was happy about the news, it was obvious that she still felt discontented with the arrangements. Carrying the food to the house, I assured her that everything was going to be okay. Kuka was already in, listening to his favorite reggae tunes on Ghetto FM. He came and hugged me, making me wonder the kind of weed my little brother normally smoked.

"Thanks Nafula for the bones. My boys have really enjoyed."

I did not say anything as I headed towards the kitchen to fetch the plates. I didn't know whether I was happy or disappointed that I had forgotten to spice those bones with Red Rat but somehow, I felt relieved that my brother would be able to take his dogs for a walk again. The following morning, mama helped me pack a few things and gave me one of her two handbags.

"In Nairobi, every woman walks with a handbag on her shoulders. Don't let the city people know that you are new."

Kuka woke up when I was just about to leave. Mama hugged me tightly, asking God to protect me in the strange lands. Although Kuka knew that I was supposed to leave that morning, he started behaving like he was hearing it for the first time. Startled, angry, and in disbelief, he rushed to the kennel and came out with his dogs. I don't know how he managed to do it but all of a sudden, the dogs started "mourning," wagging their tails in sadness, and although I was not nearer, I could swear that I saw tears flowing from the dog that had scared me the previous evening. Their master was also wailing, mumbling how his only sister was leaving him. He kept on saying that I should forgive him and his dogs if they had ever wronged me. I called him and hugged him, telling him to be good to mama.

As promised, Miss Njagi came to pick me up when I reached Nairobi. I called her using mama's phone which she had insisted that I carry along. It was around 6:30 pm when I arrived and Miss Njagi booked me in lodging.

"*Nataka upumzike vizuri*. Tomorrow is going to be a busy day," she said as she took me to my room.

"*Asante sana*," I responded, feeling that I really needed that rest after travelling for more than 9 hours. There was a small television set in the room but I could not watch, even if I wanted to. I was overwhelmed with sleep the moment I laid on the bed. I woke up the following morning and took a shower. As I was dressing, wondering what next, someone knocked. I opened the door and there was Miss Njagi with a black paper bag in her hands.

"*Ulilalaje*?" she asked as she locked the door.

"*Nililala vizuri, asante*," I responded as I buttoned my blouse well to conceal my protruding bra.

"*Nimekuletea* breakfast. Today we have a lot to do because you are flying tomorrow to Saudi Arabia."

"*Asante kwa* breakfast," I said, trying to feel courteous.

I took the milk and bread in silence as Miss Njagi focused on her phone, making long calls. She explained to me some of the activities which we were supposed to do, most importantly, processing my visa and passport. By the time we came back to my room, I was feeling exhausted and, after taking supper, fell asleep immediately.

When Miss Njagi knocked on my door the following morning, I was as ready as a lioness that had just been tipped that a troop of gazelles might be passing by shortly. I was already dressed in the *Hijab* as instructed and had the clothes she had bought me packed in the new bag. The drive to the airport was swift. Although I was anxious about boarding the plane, that did not rob away the ecstasy that overwhelmed me for riding a taxi for the first time in my life. It was around 5:20 in the morning when we reached our destination. With powerful lights gleaming over the place, nothing could have prepared my imagination that an airport could be this gargantuan and glitzy. I had always thought of airports as football fields without posts, where airplanes could land and take off. I had never envisaged that it could be a place with endless runways and flamboyant buildings. After waiting for around thirty minutes in the lounge, Miss Njagi presented my documents for verification. When the announcer called for all passengers departing to Saudi Arabia to board the plane, Miss Njagi hugged me like a mother would do to her baby and told me that somebody would be waiting for me when I reach there. She assured me that she had arranged for me to start my job at a good hotel immediately after arriving.

When I boarded the plane, I was not as excited as I thought I would be. The aircraft skid through the runway as it took off and I was surprised to find that I did not feel any discomfort. Miss Njagi had warned me that for the first timers, it was almost inevitable that they should feel nauseated, sometimes to the point of vomiting. I really waited for that moment but it did not come. Just as soon as the powerful machine started gliding through the air, my mind wandered back home and I started imagining what mama and Kuka could be doing at that moment. Now that I was not around, mama could be milking Namukhula or if she had already done that, preparing to take the milk to Eshianda-Mushimuli shopping centre. But still, she could be washing utensils or even warming water. As for Kuka, it was obvious that he was at the kennel doing this or that with his dogs. I smiled nostalgically when I remembered the dogs mourning religiously at my departure. "That boy!" I thought. I had already started missing him.

There was a time the man sitting next to me tapped me and tried to initiate a conversation but when he realized my disinterest, apologized and continued reading his book. For the first time since we had departed, I turned my head and started looking at my surrounding. The two ladies on the other side were also wearing *Hijab* just like me. I wondered what was taking them to Saudi Arabia but maybe they were also going to work in hotels. Miss Njagi had told me that there were as many hotels and restaurants in Saudi Arabia as there were commodity shops in Nairobi. I wondered what kind of work was waiting for me in the hotel I would be posted. But it was the salary of Ksh 20,000 a month that amazed me most. What could one possibly do with that kind of money? I started thinking of what I could do once I get back to Eshianda-Mushimuli. I would start by building my mother a big brick house and then buy her a car. I would then pay for my college fees and help mama to oversee Kuka through high school. The rest of the money I would channel towards business. I would also build Namukhula and her baby a nice cowshed and yes, I would also expand the dogs' kennel. It was pathetic to see six adult dogs squeezing in a tiny roofless structure.

We touched down at King Faud International Airport in Riyadh after several hours in the air. Just as Miss Njagi had promised, I found somebody waiting for me. I was able to identify her because she was holding a tag with my name on it. As I approached her, I tried to estimate her age but I couldn't. I wasn't good at such things.

"Esther Nafoola?" she asked as I approached her, pronouncing my name like someone chewing hot groundnuts.

"Yes," I replied. She hugged me tightly as if she had just found her long lost sister.

"I am Salma," she said as she guided me to a waiting car and introduced me to another lady in the driver's seat. We drove around Riyadh city, bypassing magnificent buildings with Arabic designs. Although I stuck my eyes out throughout the journey, what I was seeing was a lot to take for a naïve and primitive girl from Eshianda-Mushimuli. Even with the city behind, there were still some glamorous sites along the road as the car sped off to a new direction. After a long drive, we finally reached our destination as the car turned to a green gate, only stopping for the gate to be opened. Inside the house, there were more than 15 people, some eating, others watching television, others just sitting doing I don't know what, while the small children caterwauled and loitered aimlessly from chair to floor, floor to chair. Amidst the chaotic atmosphere, Salma introduced me to others but her words were swallowed in the cacophonous ambiance. She led me to another room and I was happy to escape from the mess in that living room.

"This will be your bedroom," she said, the Arabic accent too obvious.

"Thank you." I responded, placing my suitcase down.

"You can rest," she replied. "I will bring you supper here."

"I am grateful," I said. Salma seemed a nice and compassionate woman. "And," I interjected as she was just about to leave. "Where is the hotel I am supposed to work in?"

"It is here!" She replied, sitting down on the bed beside me. "You will be helping in cleaning the 18 rooms in the house every day. We also have a garden full of dates outside, which I will teach you how to pick. Do you like dates?"

I didn't know how to react or feel but obviously, something was amiss. Miss Njagi had assured me that I would be working in a hotel. Disregarding Salma's question, I protested,

"How…that is not what I was told by Miss Njagi."

"And cleaning the children also. They are messy sometimes but they are nice." Salma responded smiling, ignoring my protest. She was behaving like those houseflies mama could chase away from the food pot only to land at her hand, assuring her that despite their dipping in the soup, all was well.

"But that is not hotel work," I quietly interjected, feeling furious and disappointed at the same time.

Salma placed her hand around my shoulder. I didn't know whether she was trying to be friendly or the purpose she was trying to accomplish.

"Don't worry Nafoola," she said. There is plenty of hotel work. You will do cooking and washing utensils and serving the family. Real hotel work. Are you happy now?" she asked, her eyes so genuine and innocent as if she had just asked me if I was feeling well after a long journey.

Although there was little difference between the work Salma had just described and what I used to do at home, I still felt let down. But a series of thoughts about paying my school fees, helping mama and Kuka clouded my frustration. After all, mama had groomed me for hard work as I would start my morning by milking the cow and then sweeping the compound and washing clothes, spend the better part of the day doing farm work, and then finish off with lighter duties such as fetching water and milking the cow again.

Before Salma left, she pointed at a pamphlet on the bed and said that it contained all information of how I would be paid. With the door locked, I took the pamphlet and started going through it. It only had two pages. The first page contained a contract agreement of three years, already signed by Miss Mary Njagi, with the title Agent at the bottom. However, what left me shocked was the beginning of the first paragraph on the second page. "Half of the salary will be paid to the agent for the first two years in compensation of…" I threw the pamphlet on the small table inside the room and wrapped myself in the blanket. I just wanted to see if I could catch some sleep and dream about mama and Namukhula, and yes, about Kuka and his dogs. I wouldn't mind if they crept behind my back this time round.

NOTE

1 The two lines come from a common song among the Luhya people of Kenya, Namukhula being the name of the cow in the story. *Tsinuni* means sesame or simsim while *tsimbande* is a type of groundnut.

Wherever Something Stands Something Else Must Stand Beside It

A. ONIPEDE HOLLIST

The midwife maneuvered the baby out of the birth canal, spied between its legs, and gasped. "Call the chief!" she said. Hearing the midwife's sharp voice, the assistant settled the calabash of warm water on the floor and shot out of the bedroom, damp and earthy with the odor of withered leaves and herbs. Matching the assistant's haste, the midwife, who was also the newborn's grandmother, cut the umbilical cord. Then she wiped the newborn, swaddled it in a mud cloth, and held it up for its mother, her daughter-in-law, to see.

"Control yourself," the grandmother said.

"Will she be able to have children?" the mother asked, her head spinning.

"He will … but we must – "

Approaching voices drowned out the grandmother's reply. She knotted her face: "Leave everything to me." She handed the baby to its mother and pushed open the door. Waiting under a bluish-pink morning sky was her son, the Chief of the Temdogo, a people of the northern Sahel. Next to him stood his two most trusted advisors, an odd couple with wrinkled foreheads fondly referred to as Tall, the one with a battalion of teeth in his mouth, and Short, the one with only outposts in his.

"We have your successor," Tall said. "Temdogo's future is assured, unlike that of those Kridogo mosquitoes. Their king has many children, but their succession rules are so strange that war is sure to erupt when he dies." The three men chuckled. A long-standing dispute over who owned a stretch of semi-arid grassland between the Temdogos and their southern Kridogo neighbors had turned the two peoples into bitter-in-the-throat enemies. The Temdogos claimed that back when the world itself was a newborn, they had loaned the land to hapless Kridogo wanderers who now will not give it back. For their part, the Kridogos said the land had been theirs since Atala the Supreme created the Sahel: "You cannot give away that which never belonged to you" was their refrain to Temdogo claims of ownership of this borderland. And that was how two founding stories caused Kridogos and Temdogos to fight and kill each other for generations. When they could not shed blood, putdowns of the other soothed their anger. However, the birth of a Temdogo heir called for a different balm:

"Palm wine for everyone," the Temdogo chief ordered.

"Wait," the grandmother said. "I must take them to Timbuktu to see the renowned healer Alpha Khan. The mother lost a lot of blood, and we must protect the child from the evil Kridogos."

That goal motivated Chief: "Prepare to take my son and his mother to Timbuktu." Bodies scrambled. By mid-morning, grandmother, mother, and the child they agreed would be called Tamba had set off. That evening was also the first-time people of the borderland area saw a hooded figure traipsing the disputed land. They called it Dunibera, the changeling.

Sadly, the child's mother died before the party reached Timbuktu. As if to make up for not having the opportunity to save Tamba's mother, Alpha Khan put his legendary medicinal skills to work. To shield Tamba from evil, he fortified the child's body with the spells and incantations of pre-human bird sounds and Timbuktu's ancient scriptures. To assure the child would become a great ruler, Alpha Khan made small chest incisions and pressed the scriptures' black ink into them. The famous healer also fed the child monkey brain mixed with fish eyeballs to deepen his voice, grow his manhood, and make bountiful the hairs on his chest, arms, and legs. Then, he rubbed a stinging paste made from the balls of a rhinoceros and chicken gizzards to shrink the excess flesh and seal the divided that was his scrotum. Afterward, he took him to the stream, where he showed him the bodily features that distinguished men from women and explained their roles in childbirth. Of course, Alpha Khan's work took many years to protect, trying one potion and then another when they did not work, increasing their amounts, bitterness, or sweetness when they did. Many years after he first arrived in Timbuktu, Tamba returned to Temdogo. He stood at the cusp of adulthood, one of many thresholds he crossed.

In the mornings, he told stories to the children that made their eyes pop. In the afternoons, he revealed insights about life to his father and his friends that stretched their minds. In the evenings, he massaged the creaky joints of the elderly so that they could dance to the beat of the donkey-hide drums. Some days he gathered firewood and carried water with his sisters. Other days, he helped his mother cook vegetable soup. He was equally good hunting with his age mates, welding trinkets with the blacksmith, and mixing medicinal leaves with the herbalist.

The Temdogo had raised precocious children before, but Tamba set himself apart, even from adults of high achievement, because he did what many were afraid to do: he crossed into the disputed borderland and had talked, some said even laughed, with the hooded Dunibera, who was seen only at dusk and dawn. That was because, others said, Dunibera was like an albino and could not stand the brightness of sunlight. It was a creature of shades.

Always covered from head to toe, Dunibera displayed none of the outward marks that allowed one to say it belonged to this or that people. Not surprisingly, some said it was half-human half-spirit; others that it was half-adult half-child; and still others said it was half-man half-woman. Most suspected that it was half-Kridogo and half-Temdogo and that its mother had abandoned it in the borderland because it was the product of a forbidden union between two enemies. Ugh! Indeed, Dunibera's forlorn state provided all the evidence Temdogos and Kridogos needed to be sure that unity between them was intolerable, unacceptable, and unnatural.

Hotheaded youth from both peoples thought Dunibera was cursed. They pelted it with orange and banana peels. They may have done even worse had some elders not publicly cautioned that the Oracle at Tenzug Hills had prophesied that the enmity between the Kridogos and the Temdogos would end only if they rescued Dunibera from the borderland and gave him a home.

* * *

Over time, everyone in Temdogo, even the hotheaded youth, sensed Tamba's affinity for Dunibera was more than they understood. Everyone, that is, except the Temdogo chief who, one moonlit evening, his eyes liquid and his brain full of palm wine, hiccupped his worries into words:

"My son should know how to wield a sword, drink blood from a human skull, bed a woman till she begs for mercy, and command his subjects to do his bidding."

"Yes, not squeal like someone has grabbed his balls," Short chuckled.

The men snorted and then fell silent. Their worry about Tamba's manliness, and therefore his fitness to be chief, had jumped, like a virus, from the safe confines of thoughts into the volatile world of words. A troubling picture of Temdogo's future surfaced, forebodingly, like a dead animal with its entrails exposed in an early morning village square. Chief had spoken a possibility into being no one wanted to confront. Unable to understand, let alone figure a way to deal with Tamba, Tall and Short strapped on their leather bags and left.

The Kridogos and Temdogos were also unable to fathom and deal with the great famine that struck the Sahel. As food disappeared, concerns about Tamba gave way to pangs of hunger. Mixed with years of deep-in-the-throat bitterness, the neighboring peoples blamed each other for the drought. The combination erupted into three days of cross-border fighting. The bloodshed stopped only after Temdogos and Kridogos had served up each other's kidneys, livers, tongues, breasts, thighs, and legs to the underworld demons. No one alive could remember so many killed at one time. Both peoples blamed the high death toll on the guns the other had used. Both swore that the next time they would defend their land until the last bullet or breath.

It was on one killing field that the Temdogo chief stumbled on his son and a Kridogo boy weaving bracelets with palm fronds: "Are you mad?" he wagged his staff at Tamba. "Playing little girl's games with that mosquito? Drive him away," he said and held out the wooden staff to Tamba.

The son refused to take it and turned to leave. Vaap! The knobby end of the staff slammed into Tamba's shoulder. "Are you a girl?" Vaap, vaap! "Do you not have balls between your legs?" Vaap, vaap, vaap! Tamba absorbed the blows without a sound.

The Kridogo boy watched the walloping, partly in surprise, partly in solidarity. But his camaraderie reached its limit when Chief raised his staff, intent on delivering the blows Tamba had refused to deal. The Kridogo boy ran away, but not before the hate virus had entered his bloodstream.

So it was that many years after the disputed land had become arid, withered like an old man on his deathbed, the Temdogos and Kridogos still told new stories that poisoned the hearts and minds of their children. That is why in the Sahel, people say that Peace left the world when humans learned to tell stories. Today, everyone knows that if a Temdogo supported the Sahel's Peoples Party, a Kridogo will no doubt support the Democratic People's Congress. If a Temdogo started a tale about the blinding sun at high noon, for sure, a Kridogo would interrupt and say, "no, it was the moon with a dull glow at midnight."

Indeed, as the Kridogos tell it, under one such indigo sky, Temdogo snakes ambushed and killed the Kridogo king's eldest son. Everyone, including children who could barely talk, knew that the Kridogos would avenge the murder by killing Tamba. To prevent this, Chief sent Tamba – now a tall, broad-shouldered man with a long torso, supple wrists, dexterous fingers with long, clean nails – to Kumbaya. In this faraway land, old and young, literate and illiterate, believer and nonbeliever, rich and poor, able and disabled, men who married other men, and women who married other women lived in peace. This tolerance did not mean they always agreed. After all, even Teeth sometimes bite Tongue. However, unlike the Temdogos and Kridogos, Kumbaya's leaders had learned that adults must take care of children so that they grow good teeth. Because when they lose theirs, the children will be there to chew for them.

Among the Kumbayas, Tamba thrived, expressing himself with the gusto of a chameleon in a rainforest. Among the Kumbayas, he claimed Dunibera—that part of himself he had been taught to ignore, suppress. They became one, inseparable from that day forward, and asked to be called Tamba-Dunibera. They wanted to return home to Temdogo, but they knew it would not be ready to accept them. The people would have many questions, the answers to those questions would be long, hard to understand, and the upheaval would be too much for Chief. So, for many years, they made Kumbaya their home, until the day the news broke that the Temdogos and Kridogos had agreed to share the disputed borderland.

Tamba-Dunibera decided to return home. If bitter-in-the-throat enemies could overcome centuries of hate, surely they would accept them. The only agenda they had for their return was to tell the truth, reveal themselves, and become part of a new Sahel. The two peoples planned to seal their ancient division in a historic reconciliation celebration with the theme: Where something stands, something else must stand beside it. Tamba-Dunibera planned to stand beside the United People of the Sahel.

* * *

The ceremony to mark the union took place on a massive concrete slab, financed and built by the Chinese, over a large swath of the disputed borderland. The slab required 82 million pounds of concrete, the largest on the African continent. Environmentalists protested, dressed as trees, sacred bushes, swamps, and threatened animal species. Instead of a useless slab, they argued, government should have used the concrete to build low-cost housing for the poor. The project was yet another example of government waste because, for all the concrete poured, a six-by-eight-foot patch of earth remained uncovered in the center of the slab.

"That piece of earth symbolizes the narrow outlook of the past contained by the force of collaborative tolerance," the artist commissioned to memorialize the reconciliation defended his creation. The environmentalists rolled their eyes, grabbed their throats, and crumpled to the ground, feigning disbelief at the explanation. The politicians nodded their heads and used it as talking points.

Around the slab, sepia-toned canopies shaded the VIPs. Neighboring chiefs, priests, elders, cabinet ministers, and title holders attended, donning regalia, coats of arms, and other identity badges. Potentates, diplomats, and dignitaries worldwide attended, their agbadas, evening gowns, tuxedos, and designer dresses reflecting state colors. Stern-faced representatives from Israel and Palestine, Sudan and South Sudan, India and Pakistan attended, then vowed to leave when they found themselves sitting next to each other. They calmed down after the Sahelian organizers reminded them that where one person sits, another must also have a seat.

Ordinary Kridogos and Temdogos, who before would have avoided each other, walked side-by-side, talking, and laughing, their children skipping and giggling behind them. Inhabitants from neighboring towns attended, their bodies shiny with coconut oil, decorated with henna, or clothed in luxurious fabrics. Those who arrived early found seats in the hastily built stands. Those who came late stood up on the slab around the circular platform constructed for the Temdogo and Kridogo royal families. There was much joking, laughing, and cheering. The aroma of well-being sauntered through the gathering.

A pink wind tickled the cheeks of the assembled. Leaves and branches of Acacia shimmered. Cocks crowed. Birds tweeted and trilled. Cats and dogs, chickens and cockroaches, snakes and mice, usually mortal enemies, played with each other. Ruby-eyed leopards and hyenas sauntered in from the bush, their lacerating claws and fangs sheathed. Even the fierce Sahelian sun dulled its gaze as a sign it, too, was willing to make the reconciliation a success.

At 10:00 am sharp, yes, I said 10:00 am sharp, Chief and the Kridogo king strode side-by-side onto the platform, donning traditional headpieces and holding golden shields and spears. Following them came the priests and advisors. As VIPs, they sat in the front row of sofas. Tamba-Dunibera sat directly behind their father with their grandmother and other family members seated to their left in the second row of plastic chairs. A young child of the Kridogo king sat behind his father. Close family members sat to his right.

Once seated, an expectant hush settled over the gathering. The chief priests stood up, recited ancient incantations, poured libations, and commended the occasion to the ancestors. Then followed the speeches. Many speakers wove the reconciliation theme into their remarks. The Canadian speaker said, "where English is spoken; *on parle aussi le Français*." The speaker from New Zealand noted that "where a Pākehā stands, a Māori stands beside him."

But some speakers offered different messages. The American speaker said the Temdogos and Kridogos were making themselves great again, the North Korean that greatness was not the property of one country, the British that to achieve greatness one must be willing to exit from mediocrity, the Israeli that only a chosen people can achieve greatness, and the Indian that greatness without Kashmir was hollow.

The highlight of the reconciliation ceremony came as darkening skies obscured the sun. Chief stood up and stomped toward the microphone to the Dooammm, Dooammm of the donkey-skin drum. Then, in a voice so powerful that it broke into the abode of the ancestors, he pronounced: "Where a Temdogo man stands!" He stopped. Anticipation swooped through the gathering as it waited for him to complete the reconciliation refrain. He did not. Instead, the Kridogo king stood up and stomped forward until he stopped adjacent to his peer and boomed into the microphone: "A Kridogo man stands beside him."

Cheers, whistles, and ululations erupted from the crowd as the leaders danced in unison, four steps forward, four steps back, each step punctuated by the reconciliation theme's call and response. Two bodies transformed into one choreographed movement, one tale. It took only a few seconds for the gathering to catch on: "Where Temdogos Stand, Kridogos must stand beside them," a massive wave of call and response rolled over the borderland. Thousands and thousands of feet trampled the ancient bitterness into the concrete slab, sending its congealed pain down into the Underworld dwelling of Ravana, Lucifer, and multi-headed malevolent spirits. They huddled together, fearful that the humans, among whom they had sown discord from the beginning of time, had risen to exert vengeance. Meanwhile, Upper World gods and their avatars – Rama, Utanapishtim, Jesus, Krishna, Mohamed, Ogun – smiled. They had never seen humans so unified. Indeed, world media reports proclaimed that no other gathering in human history matched, or would ever match, the day the Kridogo and Temdogo assembled in reconciliation.

Tamba-Dunibera swelled with joy at the sight of their father and the Kridogo king awash in acceptance from his seat. Once purveyors of intransigent hate, the two men basked in their ability to achieve cooperation and unity.

"Where one thing stands ...," the Temdogo chief restarted the refrain for the umpteenth time—

"... we must also stand between you," Tamba-Dunibera said, thrusting apart the two leaders like an explorer wading through dense jungle foliage. They grabbed the microphone, and repeated in frenzied happiness, "we must stand between them, we must stand between, we must stand between them."

Silence swept over the gathering. No one understood what Tamba-Dunibera meant or what the tableau of them standing between the two rulers, transfixed in incomprehension and alarm, symbolized. Nature gave them no time to make sense of what they had just witnessed. The skies unleashed rain, thunder, and lightning and dispersed the gathering. Barely an hour later, the slab laid bare in a dark, watery borderland.

Many believed that the ancestors had sent the rains to wash away their bad blood, to remind them where the rain began to beat them. As a rainbow formed in the failing day's light, many would later say that Dunibera was never seen again after the reconciliation.

* * *

Chief, his trusted advisors, the grandmother, and Tamba-Dunibera dried their bodies in his living room as night closed the historic day. Chief opened his old Grundig Gramophone set and pulled out a bottle of Gordon's Gin from the space the turntable once occupied. Tall and Short snorted and pulled out their drinking cups. Tamba-Dunibera and his grandmother drank water.

"The first thing we need to do is to find you a wife," Chief said.

"We don't want a wife," Tamba-Dunibera replied.

"You will need one to give you a successor," Short said.

"To exercise your manhood. If you don't, it will shrink, you know," Tall smiled.

"Yes, but..." Tamba-Dunibera began, then paused and took a deep breath.

"Are you lesbian?" Short pounced into the pause.

"No, we are not."

"Are you gay?" Tall followed.

"No, we are not."

"Why do you keep saying "we"? There is only one of you," Short frowned.

Tamba-Dunibera exhaled. "Because I am neither man nor woman ... but both."

The revelation settled on them like lime on the tongue—sharp, sour, unpalatable. Chief and his advisors turned to the grandmother, their eyes feral with incomprehension.

She nodded but said nothing.

"What does that mean?" Chief asked. Do you have a disease?" He rose from his chair, pushed, poked, shook, and checked the locks on the doors and windows.

"He is both," the grandmother said.

"But what does that mean?"

"It means he was born a dunibera!"

The word broke out like flatulence in polite company.

"How can that be?" Tall's hand performed an instinctive reconnaissance of his crotch. Everyone noticed but no one laughed.

"But what does it mean?" Chief stood up, his shoulders slumped.

"What do you mean, what does this mean?" Tamba-Dunibera said, staring at his father, the outlines of a scowl mushrooming on his face.

"I mean you are not..." Chief's finger half rose toward Tamba-Dunibera but veered right and downward to the floor. "This is not normal."

"Those Kumbaya people have done something to you," Short said.

"He was born like this," the grandmother said. "That was why I took him to Alpha Khan in Timbuktu. He treated him with herbs and other medicines. When that did not work, he trained him to be a man."

"But what does that mean?" the Chief said. "Could your mother not have given me a normal child?"

"We are normal."

"How can you be when you have...? "

"We are."

"The important thing is that he looks like a man."

"But I am not a man."

"But you have to be one."

"No, we want to be me."

"But how can that be?"

"How? Do Temdogos and Kridogos not want to be one people, standing by each other?" Tamba-Dunibera said.

"You, he, she, they, we. All of you go back to Kumbaya! Temdogo will chart its course."

"Have you not learned anything from today," Tamba-Dunibera said. "Can you not stand beside a child of this soil?"

"Stand? The question is, who will lie under you to give Temdogo its next chief?"

"I don't know. Perhaps your son. Maybe his daughter." Tamba-Dunibera pointed to both advisors in turn.

The prospective scenarios surprised then alarmed the advisors.

"That is not the way we do things here."

"Not the way we do things," Tamba-Dunibera's voice curdled in mockery. "If you had said that to the Kridogos, the reconciliation would never have happened."

"True, but we stood by the Kridogos because doing so solved our problems. What do we solve from standing beside you?"

"Can you not do something because it is the right thing to do? Can you not change for progress?"

"The kind that threatens our way of life. Didn't our troubles with the Kridogos start after Kumbaya people told us to build fences and put our thumbs on papers which they said were marks of progress?"

"Can you impregnate a woman?"

"Can you carry a child?"

"What kind of questions are these?"

"You tell us that you are both a man and a woman, and our questions surprise you?"

The conversation continued deep into the night as Chief, his advisors, and grandmother struggled to make sense of Tamba-Dunibera. Their revelation had changed the landscape. The familiar sandy brown, rock-strewn soils, the shrubs, the dust whirls, and ashen heat of the Sahel they knew had disappeared and was replaced by a prickly humidity, a shapeless, blank, undefined territory. This new space frightened them. No one knew how to own it, but it was real estate they had to possess, one they could not afford to let fall into neglect and dispute. They knew only that by morning, they needed to have found the words to tell a new story that would keep the unity and the peace that had come back to the Sahel, nature's transitional zone.

The Song-Warrior

REGINALD OFODILE

She awoke with a song in her heart. Her elation would not be confined to her soul. She opened her mouth and warbled: 'He has given me victory...'

'Wenie,' her husband jested, 'can't you wait till we get to church?'

'The Lord is everywhere,' Wenie declared. 'Wherever we are is a church, for He is everywhere.'

'You got into the wrong profession,' the man joked. 'You should have joined a band.'

'And, Poro, could you have married a woman who toured with a band?' He was silent, and she teased, 'Africa Man!'

'There's this couple I used to know,' Poro told her. 'The wife had a really fantastic voice... soprano. When she sang, I had goose pimples... Then, once at the University Arts Theatre, she came and sang at some cultural event. Really brought the house down. People stood, cheering, clapping. After all that, she came and sat beside her husband for the rest of the events. I was on his other side. You know what he told her, in that hour of her glory? "This is the last time you're getting up to sing in public. You forget you're someone's wife. You won't make me lose people's respect."'

'Africa Men!' she chuckled, and resumed her crooning. That morning's euphoria could not be suppressed.

Wenie attended to the morning's routine – supervising the sweeping by her house-girls, their serving of a light breakfast, preparations for lunch – then her own bathing and dressing. She sat beside her husband in their Escalope jeep as he drove to the day's most important engagement: the church service. On Sundays, they dispensed with their driver. It delighted Wenie to sit beside Poro as he drove. It recalled their courtship and early married life, before prosperity and status made drivers indispensable in their lives.

After church and lunch, she was relaxing in her sitting room when a car, let in by the gateman, rumbled in. She knew that her second happy engagement that day was imminent. A tall and sprightly male form emerged from the driver's seat, then went round to the other front door to open it. A woman handed him a white-draped bundle, then eased herself out. A young girl holding a large bag stepped out of the backseat.

Wenie surmised that her guests were Junior and his household. She delightedly rushed out to receive them.

Junior was the son of her school friend, Clara. He had been an enchanting toddler: witty, charmingly impudent, and affectionate. Wenie had not communicated with Clara for many years. Clara and her family had moved to America, although the break in their association was not caused by distance. They had merely drifted apart, and over 20 years had passed. One day, Clara contacted her on Facebook, requesting her phone number. They had a merry, reminiscent chat, and Wenie was glad to learn that Junior had been back in Lagos for three years, working for an investment bank, married to some Nigerian girl he had met in Seattle, and, gladdening to Wenie, was the father of a 10 month old baby.

Wenie sat with her guests in her sitting-room. 'My husband doesn't toy with his Sunday siesta,' she explained. 'He's frantically busy all week, so this is the day when he rests.'

She teased, 'Junior, the things you used to say. We called my daughter Bilo your girlfriend. Whenever your mother scolded you, you'd go crying to hug and kiss Bilo... and my husband said, "that Junior, once he is 10 years old, he'll stop coming to my house."'

Junior chuckled, obviously unable to remember his gimmicks as a toddler, 28 years earlier.

'After Bilo,' Wenie said, 'I had two boys, then a girl. Alono and his brother are in the US. Gloria is in England, the only one yet to graduate.'

She appraised Junior's wife Vivian. Vivian was clad in subdued elegance, exuding decorum. Wenie learned she was an engineer, a graduate of the famed MIT (Massachusetts Institute of Technology). She enquired about Vivian's family, and discovered that they were people she knew, respectable and well-off.

'I'm so pleased,' she told Junior. 'Vivian is the sort of girl you should marry. I know why I'm saying this. You know, overseas is a leveler. Some of our people there marry those they wouldn't have married in Nigeria. It's easy, overseas, to forget background, but it's very important to know the lineage you're marrying into. This Western idea, "you're only marrying an individual"… it's just not true. You marry a family, a lineage.'

Wenie silently reflected on how like her mother she had begun to sound. When she was a student in England, she had fallen in love with a fellow Nigerian from another ethnic group, about a hundred miles from her own hometown. 'What?' her mother had raged, 'you stumbled on him in England, he swept you off your feet, and now you think of marrying him. What do you know of his people? We don't even speak their language.'

'Mummy,' Wenie had protested, 'we all speak English. A woman of your education saying that?'

'If you married him,' her mother insisted, 'you wouldn't be marrying just this educated, London-trained young man. You'd be marrying his entire community, with all their prejudices, their anger that you carried off this "catch" when they have their own spinsters seeking husbands…'

Wenie later returned to Nigeria, and married a man with whom she spoke the same language.

As she dispensed lunch to Junior and his family, her heart again overflowed with thankfulness for the Lord's goodness to her friend, Clara. A young family was gladdening to her. She had not analysed her view, but it was probably borne of the conviction that a young family symbolized the preservation of values, propriety, and hope for the future. A notion that had always outraged her sprang to her mind, and she blurted it out. 'I'm glad being overseas didn't change you… There's this nonsense now about man-to-man marriage, gay rights… Abomination! I lost all respect for Barrack Obama when he began to support such evil. Like flying in the face of the Lord, going against nature.'

Junior swallowed the piece of chicken he had been munching, to say 'I've no problems with gays. I've great gay friends. My take is "live and let live."' Wenie felt Junior was spouting liberal drivel and was not worth replying to. Her phone rang, and she saw the caller's name. It was someone she suspected of practicing the vileness which Junior had just excused. That latest caller was also expected for lunch. In his habitual cloying courtesy, he was calling to say he had encountered a traffic deadlock, and would be delayed in reaching his sister's on Victoria Island.

'That's all right, Aka,' Wenie said, 'I'm at home. Get here when you can.'

Wenie's glee, which had promised to be daylong, vanished. Her brother Aka was a choking ball of *foo-foo* in her throat that no *ogbono* soup could push down her esophagus. The presence of Junior and his family suddenly distressed Wenie. They reminded her of the grievous deficiency in her brother's life, the respectability and fulfillment his own willfulness denied him.

Aka was a middle-aged, well-educated man, an artist of sorts. Many considered him brilliant, their effusions about him wounding Wenie like dagger-thrusts. What use was a man like Aka? Indeed, could he claim to be a man? He was well-educated and well-travelled, but insubstantial. Aka was wifeless, cashless and childless, and incredibly, contented. He seemed to abjure status and security, his nauseating life trailed by disquieting rumours. She could not understand him.

Wenie had striven to learn if Aka was on drugs. She found that he never smoked even cigarettes. She tried to discover if he was alcoholic, but found that his drinking was negligible. He occasionally drank a bottle of beer – never more than one on any occasion – and never exceeded two or three bottles in a week. She then concluded that he was being assailed by one of two forces: he was either insane, or was under some magic spell. Of the first possibility, she reflected that neither their paternal nor maternal lineage featured madness. She considered that the second possibility might be true of Aka, although their parents had raised them to despise such superstition.

On account of her reflections, she was actually glad to bid Junior and his family goodbye. She saw them off to their car. A startling spectacle awaited her. A man was standing before a flower bush, with his back to them. He was seemingly in a trance, appearing to worship the flowers. Their chatter roused him from his reverie. He turned to meet his sister's gaze.

'When did you arrive, Aka?' Wenie asked.

'Just a few minutes ago,' he answered. 'I was held captive by that glory.' He gestured at a bell-shaped flower with white petals sprouting from a maroon base. He exchanged pleasantries with Wenie's guests, and they left. Aka's raptures were not yet spent. 'That bloom... Do you know its name?' Wenie told him that she did not. She wondered if Aka expected her to usurp her gardener's functions, then concluded that the gardener would not know the flower's name. The man was employed to sweep and nurture, not to give lessons on botany.

'I want to paint it,' Aka enthused. 'Wish I'd brought my work stuff. You know, John Constable, Renoir, Mamet, Van Gogh all painted famous flower pictures. But, John Constable's is more accessible, because of Van Gogh's impressionism. There's something to be said for what people dismiss as 'bland naturalism' or 'merely representational.' But, when you paint those petals...'

Here we go, Wenie agonized silently. *Why is this Aka obsessed with useless information? Can't he talk about buying land to build on Lekki? Can't he talk about finding a girl to marry, and having a couple of children?*

'I never heard you arrive,' she said. 'You didn't come in your car?'

'No, it's off the road,' he told her, and she asked him how he came to her house. 'Oh, quite a journey from Iba. I took a bus to Iyan-Oba, then another to CMS, then one of the small buses to VI. I got a bike to your gate.'

Before he finished that recital, tears were rising to Wenie's eyes. Aka obviously could not afford to travel in a Lagos yellow taxi or an Uber cab. Either conveyance would have brought him to her house in one unbroken ride. A man who had been started superbly in life, chose to live a hard-up existence in Lagos. Wenie's husband, a man of high and extensive contacts, had secured Aka a plum job. Aka declined it, saying it would leave him with insufficient time for his work. Wenie had been incensed. 'What work are you doing? Messing about with brushes and paints like a child. Work indeed. Work that hasn't given you money or any position in life, and you'll soon get to retirement age.' On another occasion, she had raged at him, 'You're wifeless, childless, cashless, and you're content?'

Today, when they entered the house, she instructed her maid to serve him lunch in the dining room. When Aka went in to eat, Wenie rushed into the lavatory, locked herself in, and bared her soul to the Almighty. 'Please, the great I am that I am, the omniscient, the omnipotent, please be at my discussions with Aka. He's obviously lost his mind, and needs you… He is blind, Lord, he needs you to open his eyes…' She paused, reflecting that all her other siblings had given up on Aka, with his life of drifting and penury. Wenie however, worshipped a Divine that never gave up on his children. Her God had transformed a disreputable man who clambered up a Sycamore tree to see the Saviour. Her God had caused the repentance of a robber on the brink of execution. He, the Saviour, had even resurrected the dead. Aka's perverseness was trivial beside the villainies the Almighty had rectified. Wenie prayed further, 'Dear Lord, please help me. Let me not lose my temper today. Help our conversation to bear fruit… Please help Aka to see reason. He has his good qualities… Please do not abandon him.'

Thirty minutes later, she and Aka were discussing 'settling down.' 'I guess it has its roles,' Aka conceded, 'but it's not the only way. However you define it. We drive through Lokoja today, and remember Mungo Park, because he never settled down. He lived a wandering life, what some would call a vagabond life.'

'What about being stable?' she urged, 'having a family.'

'You mean I must have a wife and children to make me complete? No. Marriage has its uses, but it's simply not for everyone. It's not a *sine qua non*.

'Your parents got married and produced you,' Wenie gently reminded him, and he replied, 'Yes, I know I'm the product of a marriage that has lasted decades, but I'm not our Daddy. Whoever thinks being unmarried is a stigma, has an inferior mind to mine. I won't be directed by my inferiors.'

Wenie paused, and spoke again. 'You saw Junior, son of my friend, Clara. You held him as a baby. You were already an undergraduate when he was born. Now, he is a family man, a solid person, and you… you like to remain a tramp.

'Of course you have no money… You're not living. You barely exist.'

'The times are biting,' Aka conceded, 'but we can't all make money. Many of those who advanced the world were not rich. Some were even called mad. We take air travel for granted today. For decades, those inventors of the aeroplane, Wilbur and Orville Wright and Alberto Santos Dumont, were seen as people flying in the face of the Lord. Many people said, "If the Almighty had meant us to fly, he would have given us wings." Who was the richest man in the time of Socrates? Who was the richest man in Shakespeare's England, in Leonardo da Vinci's Florence? We don't remember them, but we'll never forget the Mona Lisa and we'll never forget Shakespeare.'

Wenie observed that Aka was at least spruce on that Sunday, which was not always his wont. To her, neatness and grooming were not only second to piety, but sometimes ranked above devoutness. Scruffiness and dirt revolted her. She found the trend of ripped jeans incomprehensible. How could anyone be voluntarily ragged?

Only days earlier, Wenie had received an unsettling phone call from her friend, Ovie. 'Wenie, I don't know whether to tell you this, but we go back a long way. I was passing Ojuelegba… there was a pile of *okrika* on the ground, and I saw someone like our brother Aka buying *okrika*. I don't know… I'm sure it was Aka. I thought, "doesn't he know whose son he is? Doesn't he know where he comes from?" If Aka wears *okrika*, who would wear Yves St Laurent and Balenciaga?'

Wenie, looking at Aka, felt Ovie had been mistaken – Aka's garb was too qualitative to be secondhand. She complimented his attire: 'you look sharp.' 'Cheap sharp!' Aka quipped, 'I bought everything from *okrika* at Ojuelegba.'

Her dismay was acute. As though to combat her misery, she recalled how exultant he had once made her. It was an occasion on which Aka proved himself worthy of his pedigree.

It had been announced that Aka would deliver a lecture at an international arts event, sponsored by a foreign embassy, on Lagos Island. The television and newspapers gave it extensive coverage, featuring profiles and interviews with Aka. Wenie's friends were continually calling her. Even her reticent husband, affected by the euphoria, pronounced, 'Aka is not a small person.'

As she heard Aka's rhapsodies on the phone, Wenie felt her world was seized by Cézanne. 'Paul Cézanne, the quintessential impressionist… he did marvels with light and water … he ranged from impressionist to abstract…' The name Cézanne meant nothing to Wenie!

Wenie wondered if her excited friends cared about – or had any knowledge of – the art of Cézanne. She swiftly surmised that her friends' euphoria was borne of their perceiving that the impending event was grand, certain to be attended by the eminent, expected to enhance Nigeria's foreign relations.

When Wenie and her husband arrived at the venue on Lagos Island, an imposing, high-pillared, white-painted colonial edifice, her soul exulted. She beheld a convergence of majestic automobiles: Toyota land cruisers, S Mercedes sedan cars, and two Rolls Royces. Some of the vehicles had purple diplomatic number plates, their bonnets adorned with flags of various nations. Ambassadors were certainly present. She shortly saw two cabinet ministers, as well as famed magnates of finance. It was incredible – and thrilling – that all those grandees had come to acclaim her brother. Outside the building was a huge banner, emblazoned with her family name and Aka's photograph. Aka was outside the hall, journalists and photographers clamouring for his attention, cameras flashing, Ambassadors and Ministers seeking to be photographed with him.

Cocktails and snacks were laid out in the spacious and impressive entrance chamber. Wenie was silently but fervently thanking the Almighty for a brother like Aka, a brother that magnified the family name before her associates. Family names meant much in her circle. Aka now made her feel superior to those whose brothers were executives of oil companies, or owned oil blocks, and large properties in Lagos and Abuja and Texas… There were the great and the good striding or sidling up to Aka. Aka's containment and courtesy were unfaltering. In her heart, Wenie continually blessed him.

The lecture proved her earlier conviction. Neither Cézanne nor his art mattered to most of the attendees. As Aka spoke, many in the audience chattered away. Although such indecorum was routine at functions, Wenie bristled at her compatriots' manners, their failure to recognize that even if bored by a speech, it was polite to feign fascination.

Aka stood at a lectern, a jug of water and a glass before him. Behind him, stretched along the back wall of the dais, was a gigantic screen. Aka's opening remarks baffled Wenie. 'The name Diana Vreeland might mean nothing to this audience. To students of 20th century style, however, Diana Vreeland for decades represented a fantastic fusion of glamour and applied art. She was the editor of *Vogue*. Diana Vreeland elevated colour to unprecedented preeminence. She acclaimed and explored that most compelling of primary hues: red. She decorated her boudoir to look like the inside of a ruby. Diana Vreeland declared that she could not abide any red with orange in it, but was repelled by orange devoid of red!

'Some reds are famous. There is Titian red. There's the red discovered – or rediscovered – at the Pompeii excavations. Red is a captivating – and most significant – hue. Red is life; it streams when a life is cut short by a metal implement. An outflow of red accompanied the birth of each of us. A stream of red marks the induction of every female into copulatory womanhood as...' Now Wenie squirmed, feeling her brother was lapsing into obscenity, but his explicitness was only momentary. Aka proceeded in decorous vein. He commanded partial raptness, the attentive and cognizant minority in the hall sometimes cheering. She heard someone gush, 'to think he's not reading this out; it's all in his head!'

Aka imparted, 'I've chosen red, that vivid, strident, captivating, symbolic hue as my launch pad into the artistic legacy of Paul Cézanne. To the uninformed, his depictments were antithetical to vividness, or even clarity. The casual observer thought him diffident, indifferent, even indolent. He was dismissed as vacuous, indecisive, inept. The seeming blurs and haze of Paul Cézanne's art, however, expressed as much vibrancy, as much unbridled vigour, as the paintings of artists who employed overtly vibrant hues. His was actually higher for its depth and subtlety...' On the large screen, the backdrop, a series of Paul Cézanne's paintings was beamed.

Wenie was not surprised that the lecture was succeeded by further feting of her brother. He introduced her to some eminent persons, some of whom remarked their facial resemblance. People even began to fawn on Wenie!

Deflation duly displaced elation. As guests were starting to leave, Wenie stood chatting to a friend outside the hall. In full view of the remaining throng, Aka stepped into the street, hailed a passing *keke*, and when it stopped, boarded it. Wenie's mortification was such that she felt simultaneously slapped in the face and kicked in the belly. She wished the lawn on which she stood would spring a hole and bury her. *Keke* was the transport of the lowly. If Aka used it whenever his car was off the road – as it often was – did he have to do so before an assembly of the nation's crème de la crème, on a day he was acclaimed beyond her dreams? She knew that most of the guests would have gladly given him a ride to whatever destination he chose. Minutes earlier, she had heard him refuse invitations to eat or drink with some of the adulatory attendees.

It was a dejected Wenie that returned to Victoria Island with her husband that dusk.

Today, on the Sunday of his visit, she heard his further apologia for his failures. 'Many immortal artists starved. Some poor people will be celebrated forever, while their rich contemporaries are forgotten.' Wenie asked Aka if he never sometimes felt the urge to be like his friends who were successful men of family, and he admitted, 'You know, I sometimes want to be like those solid guys, those big boys of Lagos. It'd be great to be "loaded,"... I've much hardship in my life... It'd be great to be respectable, what Cockneys call "all sorted"... what Americans call a "crew-cut" person. Yet, part of me despises such people... their being settled pillars of the community... I feel they're bland... commonplace. There's a phrase I coined; I love alliteration, "to proceed in predictable propriety and prosperity." Despite all I go through, I'd hate to proceed in predictable propriety and prosperity.'

That Sunday night after Aka's visit, Wenie acknowledged that she had once more failed. She descended into a reaction she had vowed that Aka would never again impel her into: sobbing. Then, fury subduing her sorrow, she prayed: 'Dear Lord, I'll never again allow that foolish boy-man to spoil my joy. You've given me a great day, Lord. I'm grateful; I'll keep rejoicing and being glad in it. Whatever my brother decides to do with his life is his business. Our parents gave him the best. If he chooses to be a tramp, that's his business. At your judgment throne, each must answer for themselves.' Wenie's day ended as it had begun: with a song.

'It is well, it is well with my soul.'

Six Poems

IQUO DIANA ABASI

ANSWERS THAT WILL NOT BE SWALLOWED

We are children once amused at tales of a snake
slithering away with 36 million fractions of commonwealth.
We were the learners once outwitted by a monkey
who disappeared with double the snake's loot.
We are the ones toiling for an irregular pittance,
overfed with sorrow, tears and blood
while those we empower earn
hundreds of times more.
We are Agatu, Logo, Guma, Nkanu;
despised children whose lives
would be worthy of concern,
if they were cattle.
We are Chibok, Baga, Dapchi, Mubi,
Buni-Yadi; abused, gone too soon.
We are the unloved, vagabond
children of an uncaring father.
We are the oppressed, seeking redress
where the accused is both judge
and executor. Confidence is a distant relative
slaughtered on the slab of injustice.
We are Jesse, Abule-ado, Abule-Egba,
unfortunate ones, set ablaze,
in our nation's wealth, yet blamed
for her shameless neglect.
We are the Odi emergency,
wiped out but not forgotten,
ants decimated by the sledgehammer
of power, our blood cries out; still.
We are Lekki, Alausa, #EndSARS
futures truncated, voices stifled

by rulers who send armed soldiers
to a conversation with children
We are ones whose massacre will not
be erased, whitewashed by the very ones
who strangle us, dim our tomorrow
while we grope amidst today's crumbs.
We are the infants overfed on
sorrow, tears and blood,
We are the hitherto 'Woke',
now awakened, demanding better.
We are the questions that will grind your teeth,
we are the meal that will choke you to surrender.
We are answers that will not be swallowed
in your divide-and-rule games,
We are the puppies of a scoundrel country,
determined to ride through myriad upheavals.
We are midwives of this unstoppable rebirth,
and we, will no longer be silenced!

WHEN A BITCH EATS HER YOUNG

Children refused to be silenced,
thus parents discarded kid gloves,
to quell the noise.
The interaction was swift;
a few hundred rounds to hush pesky
enemies, under a convenient power failure.
When did defender turn
slayer in peace time?
How did begging for better governance
and an end to brutality become a
crime punishable by death?
Does an absence of bodies
delete a massacre, *grammed* live?
Can lies purify the tarmac,
and bleach a blood-stained flag?
Will hidden evidence restore heartbeat,
or shattered face, or future
ripped to shreds by state goons?
Sterile state speeches and fantastic
discoveries may hold sway, but will
drama douse the fire of revolution,

burning in resilient hearts?
When endurance is stretched beyond snapping,
will tactical erasure and veiled threats
yet suppress the people's will?
When the oppressed break those
chains and embrace anarchy,
will the centre yet hold?
When a bitch eats her young we seek a vet,
blame hunger, post-partum afflictions.
When a scoundrel country eats her youth
who will diagnose the ailment,
incise the vein-deep decay.

THIS IS HOW

This is how we refuse to remain
citizens of a heartless country,
children of wicked forebears,
parents whose only care is loot
and perpetual power; products of
a recycle system configured like
marriage vows; till death do us part.
This is how we undo the socialisation
of silence, and reject the rot,
when enough is enough, we become
stupid in the eyes of others,
we stride into death's grip,
if only to make our voices heard,
demand change.
On our knees.
This is how we process the bloodshed.
Weak, as rulers bare fangs,
get high on priceless champagne,
their glasses spun from
the skulls of our young ones.
#EndBadGovernance #EndSARS
Unfazed through the chaos,
this is how we attempt to pull back Motherland
from the claws of our oppressors;
one-time youth who tasted power,
then chose retrogression over development,
chose to squeeze the nation

of her last juices.
This is how we march
through ruins of failed state,
armed with a cupful of tears,
two cups strategy,
two cups focus, and resilience;
determined to succeed, in honour
of those slain at the dawn of our awakening.
Lekki,
Alausa,
Ogbomosho,
Enugu…
Ozoemena!
This is how we look one another in the eye
and declare: Never Again!
#EndSARS
#EndPoliceBrutality
#EndBadGovernance

A DAUGHTER, COMING UNDONE

Dear Father,
I attempt our English grammar lessons
every evening, as the sun sets,
but syntax was long lost to the
shattering sound of gunshots,
no more concord exists,
between the echoes of this longing
and razor-sharp precision of machetes.
You were my angel
cut down for defending his own,
my hopes melted in a blinding fire
that rid me of home, hearth and peace.
I knew not where we were taken,
what nightfall had not blanketed from view,
teardrops ensured I did not see.
Still, my will refused to be broken,
impassive, no torture would undo me,
but the mind is sometimes
impotent before matter.
The body did not break, responding
to its own rhythm, it chose to swell

instead, like soil after ploughing.
I came undone, as new life blossomed in me.
I sought an exit; like Talatu who
jumped off the truck when they moved
us across camps, like Fanna who breathed
her last, unable to expel her baby.
Death refused my many supplications,
I birthed Shehu, the colour of midnight, like his father,
with curly hair and quiet eyes like you, father.
This is my certificate in lieu of SSCE,
how do I hate what I love so fiercely?
Soldiers set us free,
but we pay the price for this freedom
in a currency minted with bits of sanity.
Father, your people avoid me like contagion,
no longer your daughter,
I am touted a killer's wife,
my fruit labelled *BH gene*.
I can neither shield him nor teach confidence
when parents keep their children away.
Like fragile fabric,
coming neatly undone at the seams,
I am held apprehensive by allure
of cough syrup and tram.
Sanity is a high escape
from this stigma
gradually
squeezing
my will
to live.

CRUMBS

Rush at the crumbs,
as masters swipe them off table's edge,
unashamed, beg for more,
promise allegiance and loyalty;
anything to secure continued droppings
from the bowels of black crude
beneath your father's farm.
Woo them, saying:
'Boss, how far, you know your boy is loyal!'

'Chairman, make dem leave something for your boys na!'
Remain a faithful 'boy' while
contaminated air and polluted rivers
number your days.
Ever loyal,
do his messy cleanups,
erase unwanted elements for the master.
Do anything,
but don't engage the 'Boss' in serious
conversation about community entitlements,
oil spills, education, jobs for youth.
How long before you tire of
man-made crutches, while
your limbs remain dormant?
When will you awaken and ask
whence these handouts came?
Or query what these millions were set out for
before they turned crumb for boys?
How long before you stand,
demand a say, in the sharing
of this communal meal?
A meal for which your land's
depths were plumbed, while you
continue to live in penury and filth!
When will you see that masters
only push crumbs your way to
keep boys on a perpetual leash?
Land of immeasurable wealth turned
poster child for despoliation.
Oh, Niger Delta,
when will you tire of this bondage?

NOT CRYING

I'm not crying,
No.
These are not tears,
but my heart breaking up
into innumerable fragments,
these tear ducts are mere pathways,
the crooked road to accommodate
the relentless vessel of my love

cascading in torrents,
liquid love puddles at my feet
calcifies in meaningless shape.
these are not tears, no
believe me
I seek meaning,
Seek reason,
Drown in blood gone cold as
Icicles pierce the void
This heart needs one more tear,
One last drop of blood
One last breath…
I come up for air,
This heart must be whole again,
but I am not crying
I tell you; just washing away
the log in my eye.
these are not tears,
but my heart in fragments…

Three Poems

AISHA UMAR

THE STRING OF DISCORD

Unsettled by the peace of the world
They schemed to sow the seed of discord in our farms of living
Through chimneys of annihilation,
Vessels of separatism,
Sermons of false truth of medicine.
Our unity, they flushed with fear
Livelihood, washed away with waves of pandemic
Making us prostitutes of needs
Merchant-beggars for light and might.
As they trade our freedom for petals, sitting on the throne of deals,
The media become our feeding bottles of lies,
As the eagles shower their droppings for revitalizing and cleansing the land,
Arithmetic wears a new garment.
As the string of discord binds us in pain,
Siblings become strangers,
Friends, ceremonial enemies
Hospitals become death malls.
No embrace or hugging, they say
For embrace died among lovers,
Hugging becomes a luxury beyond the reach of even the rich.
Handshake is bedridden, awaiting death or recovery
This compound word, in our imagination transplanted,
Becomes the death knell on our family ties.
The child running errand is trapped,
And father cannot go in search.
Husbands wallow in masturbation,
Confined to their farms, away from their wives
Feeding fat from the produce of their strife.
Some sleep with their wealth under their pillows
With no one to barter with

Others are victims of hungerdemic and poverdemic.
Living has never been more worisome and death more worthy,
We ride in official vehicles to malls,
Accompanied with the blaring siren
That pleases no ear,
Awaiting death in isolation and funeral by strangers.
The iniquity of this world is the brainchild of barons
Dressed in the string of discord.

DESTINY'S DISH

Just when I thought I am at destiny's edge
The remnants of yester years came calling
Its face ugly
Voice husky and horrid
Its revelation mind blowing
You never know with humans, I say.
When over whelmed by sleazy emotions
We succumb to lusty fits
That ooze out of our pores.
The realization though choking
The absorption, unruly
The effect damaging
Leaves me pondering over the depth of the human mind
Shredded by conflicting emotions, disappointment and betrayal
I'm weary with the sting of reality
Imperfection is man's inner skin.
What does one expect,
In return for gestures served?
What did I expect from this callous world
In return for my benevolence?
Is the world even aware of my benevolence?
Does the world believe in my sincerity?
Or be contented with my selflessness?

TASHA

Like a hurricane,
Your presence swept over me
Dotting and possessing
Without caution but zeal
Defiling all warnings

Scaling all heights
Over-stepping all boundaries
My presence made you tremble
And I crumbled before yours
Worries of uncertainty
But what we share, spanned us dry
And voila!
Our beings spoke the silent language
Our eyes bore the joy of discovery
And the evidence of yester yearnings
Confused everyone else is
They all are concerned
They surely are
For a future heavy with uncertain birth
Afraid of being a catalyst to err
You have come a long way
Crossing many dry rivers
I have not travelled less
Cascading slippery mountains and narrow bridges
Broken at times we both felt, wishing to end it all
But the unknown knowledge we felt
Of what is yet to be
Powered our will
Quietly illuminates us
In the serenity of a world we both share
With no neighbours, nor folks

Poem

TIJANI ABDULLAHI OLANIYI

AFRICAN CHILDREN

On our rich African land,
Sat our children under the watch of full moon,
Enjoying the cool breeze that came
With the night.
Different cardinal point they were,
Yet bound by a future expectation.
Happy faces they wore but sad heart they had;
Present in person but absent in mind.
Living in reality but lost in dream,
Of what is and ought to.
Different dreams they had –
Some doctors, others lawyers;
Some engineers, others economists,
Some teachers, others bankers,
But they wished for the same future.
A future,
Immune from sickness and struggles,
tears and Pain,
mourning and sadness.
A future,
Wrapped with prosperity and growth,
joy and Freedom,
happiness and life.
A future filled with love,
Is all they ever wanted.
That wasn't too much to ask.
Yet the future they hoped for,
Was far away from them.
Over and over, they'd toil the land;
With joy, gathering the crumbs

That fell from their parent's table.
Nay, the soil they see is but a barren toy.
Who is going to give them this future?
Who is going to save their land?
The Giant among them has fallen, fallen faced
Down.
Alone they are
In the unending struggle
For survival.
With the sun rising,
They couldn't but hope,
That joy comes in the morning.

Poem

CLARA IJEOMA OSUJI

NUN'S TWILIGHT CALL

Riverine.
I have known riverine poetry as Okara,
Through the classic call of his River.
The jungle drums and concerto Sage
Effusions of wisdom laced with eloquence,
Craftsmaster who offers a piece of his life
To every reader; unleashing sagas of seas, creeks
Nembe River, watersides, canoes and fishermen.
Like a vestment, the Sage who wears words
As if his life and craft expressed an infinite space,
Blazed forth, enlightening the poetic cosmos
Living the lyrical existence – a good slice of life
Timeworn, yet, crafting words immortal.
Summons.
Albeit, the River had called before –
Now, she summons: insisting, urgent, final
A different call, not of serenades and reminisces
The Sage must make haste; he steps out swiftly.
See, around his writing table – an empty space
And a lonely pen too sharp to be abandoned,
The tested writing implement aversely ebbs
From hewing magical words in prose and verse
Provoking reflections of tranquil craftsmanship.
Mortality.Death, the only infallible contender
Forever plunks the frailty of mortals to test.
O irreproachable Sage!
Like a river, the human life ebbs and flows.
Who shall tell us new tales of creeks and fishermen?
Of streams, beaches, sea waves, corals and ferns?
Eloquent profusions of a life of meditation in words

Crafted to instil and defend a distinct brilliance,
Elegance tempered with frugality.
Rouses.
The Sage, laureled, full of age, glory and honour,
The greys of the moment attesting to drifted seeds of time,
Audaciously, answers the final summon of his River.
Like the ebb of the tide which comes at twilight
The Astute Guru disengages from the mortal plane
At last, the debt of Nature has been paid.
By sundown, the Irreproachable Statesman
Is exhausted, but keeps trudging on,
His twilight-self, still exuding elegance.
Traverses.
At the creek, the waters rise up in grandiose recognition,
In reverential homage to the Impeccable Sage
In fair veneration of the gentle Man of Letters,
A canoe waits by the riverside – stately, bare and quiet.
Gallantly, the Unvanquished Composer progresses
Unhurriedly, eschewing the temporality of tasteful pasts
Of life immersed in the sublime art of poesies,
The Sage sails to his ancestors on the laps of the Nun.

Poem

IFEOMA OKOYE

TO MOKWUGO OKOYE
A FORSAKEN FREEDOM FIGHTER

You were born before your time
And that is your only crime,
But are you to blame
That people know not your worth
Or fail to come forth
And tell what they know of your fame?
You spent three years in jail
To make sure they would not fail
In their freedom fight.
You shunned wealth for the war to be won.
But in sharing the spoils, none
Thought of your role in the fight.
The truth you have always spoken
Without fear of who is broken.
They named you a censor,
Lone voice in the wilderness.
They tried to ignore your goodness
And in the process made you greater.
They mocked your simple aura,
And gauged your worth in naira,
But to you it is evil
To enrich oneself and breed
From corruption and greed
And by conniving with the devil.
But I have gained where they have lost
And will keep you at all cost,
For your love is most true,
My path your knowledge has lighted,
My virtue yours has hoisted,

And my hope has sprung from you.
As our love shunned tradition
To bring itself to fruition,
So shall our goodness,
Our hope for future bright
Serve us as a powerful light
As we bask in true happiness.

Tributes

Remembering Eldred Durosimi Jones

(6 January 1925 – 21 March 2020)
Farewell, Othello's Countryman

NIYI OSUNDARE

I met Eldred Jones personally for the first time in 1982 under a circumstance I have found delightfully unforgettable. From his base in Sierra Leone he had been invited to the then University of Ife (now Obafemi Awolowo University) for a professional engagement. But before leaving home, he never forgot to tuck in his travelling bag a file containing the manuscript by a rookie academic, submitted and accepted for publication in *African Literature Today (ALT)*, Africa's influential literary journal for which he was founding editor. His engagement over in Ife, Professor Jones set out for the University of Ibadan to see the author of that manuscript so as to clarify two 'issues' with him: one, the correct version of a Yorùbá name as spelt in the manuscript, the other a debate as to whether the correct punctuation mark in one sentence in the middle of the essay should be a comma or semicolon! It took no more than a few minutes to iron out these 'gigantic' issues, and Professor Jones put back the manuscript in the file, the file back in the bag, ready to go. But he did not leave without commending the accepted manuscript both for its 'quality' and its 'usefulness'; for according to him, we needed more of its kind of stylistic, language-based study to further widen the scope of a journal that had become an authoritative platform for African literature in all its ramifications. A few months later, *African Literature Today 13* was out, with my article, 'Words of Iron, Sentences of Thunder: Soyinka's Prose Style', between its hallowed covers.

The impact of this first meeting with Professor Jones has remained with me to this day. Witty, easy-going, and personable, he was the kind of elderl/leader who put you at ease with a ready smile and sincere fellow feeling. One of Africa's renowned scholars and university administrators, he carried himself with grace shorn of showy grandeur, simplicity ennobled by superb delicacy, a near-magical aura that announced his presence in the public square without the effort of any

Master of Ceremony. A scholar and teacher with a monk-like fidelity to detail, Professor Jones taught and demonstrated that belief that no academic subject was too insignificant for the scholar's investigation; no matter too little to attract their meticulous consideration. Thus when Professor Jones sat by my side and engaged me in a finnicky differentiation between a comma and semicolon within the syntactic structure of a sentence, his lesson went way beyond the punctuational protocol of academic writing. That lesson foregrounds the ability to recognize the big in the small; the curious visibility of unseen things; the ultimate fact that big things derive their bigness from the concerted complexity of little things. That very ability is itself the product and manifestation of an attitude which bodies forth as a propensity; a meticulous method which becomes a habit. It is the kind of attitude that makes the difference between development and underdevelopment, success and failure. It is the kind of thoroughness so indigenous to old Krio culture and pedagogy in *Saroland*, the kind that many Nigerians of my generation learned from our high school teachers, some of whom were proud alumni of Fourah Bay College, Africa's oldest Western-style university institution in West Africa.

A Shakespeare scholar and acclaimed authority in English letters in all their astounding aspects, the author of *Othello's Countrymen* knew from both personal and professional experience the vital role of literature in the shaping of a people's culture, the revitalization of their memory, the cultivation and maintenance of their sense of self and racial pride. In a nutshell, he was achingly conscious of the fact that for a people to be in a position to control their own destiny, they must, of necessity, be in charge of their own story. Africa's story and the critical responses to it had, for the most part, been shaped and determined by forces beyond the continent, and the rise and evolution of written modern African literature as we know it today were supreme attempts at the redressing of a gross exogeneist and colonial disadvantage. Eldred Jones was there at the birthing of this literature, and up to the moment he breathed his last, he was one of its crucial nurturers and sustainers. This is evident in the readiness and enthusiasm with which he poured his vast erudition, insight, incomparable energy, and organizational acumen into the editing and management of *African Literature Today* from its infancy until it became one of the most authoritative platforms for the articulation of critical responses to African literature and – very important – re-orientation of its aesthetic paradigms. As Syl Cheney-Coker, the eminent African writer and Professor Jones's compatriot, has said most eloquently in his own tribute, barely five decades after its founding, *ALT* has become a thriving arena for the finest critics of African literature and fertile catalyst for the growth and development of the literature itself. Thus, Jones facilitated the creative-critical synergy so fundamental for the evolution of African literature as a balanced, sustainable enterprise.

Eldred Jones was not only an enabling force and nurturer of a burgeoning literary tradition, he was also one of the continent's leading scholars and literary figures. His work, *The Writing of Wole Soyinka* (1973) was the first book-length study of an author who later justified Jones's vision and promise by becoming the first Black African to win the Nobel Prize for Literature. But by far his most famous work is *Othello's Countrymen: The African in English Renaissance Drama* (1965), a work whose scholarly and sensitive exploration of the literary representation of the African establishes a powerful concert of insights with studies such as Es'kia Mphahlele's *The African Image* (1962), and Chinua Achebe's generally acclaimed 'An Image of Africa: Racism in Conrad's *Heart of Darkness*' (1977). And, less than a decade before his transition, he endowed his beloved city/country with an unforgettable memoir, *The Freetown Bond: A Life Under Two Flags* (2012).

A staunchly resilient, personality, Eldred Jones led a life whose triumphs were almost equalled by its travails. He was struck by blindness at the height of his professional powers, but he never allowed this adversity to dim the light of his mind or terminate his lifelong love for books and ideas. He kept up his position as chairman of the Noma Award Jury, and was a prominent resource person on the AFRICA'S BEST 100 BOOKS OF THE CENTURY Selection Committee whose award presentation took place in Cape Town in 2002. In the major literary session accompanying that event, Eldred Jones held us all spell-bound with an erudite, wide-ranging, and current lecture on African literature, even without a shred of paper anywhere around him! When the event was over and I asked him in open-mouthed astonishment, 'how did you do it, Prof.?', he held up his wife's hand and said 'Ask Marjorie; she's the author of the magic'. Marjorie's smile bloomed into radiant laughter. There was so much tenderness in that laughter; so much beauty in its 'author'. Over the years, I thought that the union of Eldred and Marjorie was made from heaven. African literature, nay the world's literary heritage, is the beneficiary of that union. For Marjorie was not just a wife; she was also a soulmate, helpmate, facilitator, most valuable enabler, and, above all, a warm and generous HUMAN being. We can never tell the full story of Eldred Jones's role in the evolution of modern African literature without a long chapter on this Angel with abundant earthly beneficence.

And so Eldred Durosimi Jones has crossed seven rivers, climbed seven mountains, and landed softly in the valley of immortal ancestors. When in May 2016 our documentary team paid him a working visit in his hilltop residence in Freetown, he was about 91 years old. But his boyish face, mellow manners, poignant erudition, and tenacious memory teased me into the illusory feeling that he was just too good to die. But as the Yorùbá say, *gbese n'iku*. Yes, death is a debt. Durosimi has paid that debt now, and his reward is unvarnished immortality. Sleep well, Othello's Countryman. Here now, a snatch from that song I sang when you turned 71 in 1996, and performed personally for you in your Freetown home some four years ago.

Live on, then, my Scholar-Poet,
Twilight shadows may be long,
who does not know that Life is stubbornly l-o-n-g-e-r?
Your vision now so internal, so eternal,
keeps capturing the bird on tomorrow's tree
Ripening rice fields dance towards your table
the raffia palm drops its honey
in the sanctity of your gourd
in your garden of a thousand fragrances
petals bloom into doves and rainbow spaces
Let rivers empty their burdens into the waiting sea
let the moon wane and wax
in the laboratory of its sky;
your liberal wisdom sews the seasons
into proverbs and silky laughters
Teacher, founder, author of humane letters,
Your drum throbs in the marketplace,
gathering metaphors and leaping visions
you who master the immensity of night,
arriving, still, with vistas of new dawns

Iyeyeluye
Alugbinrin jan ki jan ki jan alugbinrin

Professor Eldred Durosimi Jones

A Humanist and Critic

ELIZABETH L.A. KAMARA

Eldred Jones's illuminating memoir, *The Freetown Bond*, reveals a lot more about him as a man than we could have gleaned from his literary or scholarly writing. This work demonstrates that Jones is a symbol of education and humanism. To be sure, his memoir is not only about himself, but is also a book of tributes to all those that he had come into contact with who touched or enriched his life in a certain way.

Jones is a humanist who has always demonstrated the virtues of patriotism, altruism, refinement, humility and love of learning. No one will deny that these positives are slowly approaching their sell-by dates in present-day Sierra Leone. In this light, I have chosen to celebrate him and the institution he epitomises in order to inspire others to emulate a dying breed.

I have attempted to examine the humanism of Jones from the standpoint of A.C. Grayling thus:

> humanists aspire to be ethical agents who wish always to respect their fellow human beings, to like them, to honour their strivings and to sympathise with their feelings. They wish to begin every encounter, every relationship, with this attitude ... and to join with their fellows in building just and decent societies where all can have an opportunity to flourish.
>
> For that is what humanism is: it is, to repeat and insist, about the value of things human ... It is about human life. ('The Milk of Humanist Kindness')

Jones, Emeritus Professor of English and a national monument, is one of ten children born to Eldred P.W. Jones and Ethline M. Jones in Freetown on 6 January 1925. He is a bonafide member of the Freetown society in which he lives and spends almost his entire life. In their own rights his parents were also humanists, though I do not believe that they labelled themselves thus. Note how Jones describes his childhood family home:

> Our house was indeed full of children, cousins real and acquired, many of whom were sent to my mother from the peninsular villages for 'training'. At one time, five real cousins, three girls and two boys arrived from Nigeria where their parents worked, the girls remaining in our family until they married. (*The Freetown Bond*: 7)

The home described above is a typical African – especially Sierra Leonean – household, teeming with children, one's own and others who become by association children of the house. The economic factor and modern lifestyles have rendered such households no longer as common in Freetown as they used to be, but are to be found in most of the communities in the provinces, where people pride themselves on the number of children they can feed, regardless of whether they can fully meet the needs of all members of their households.

Jones's humanism has clearly been nurtured by his parents 'training' of so many children and treating them all as equals in spite of the fact that they also had a lot of biological children of their own. In those early days, his parents must have instilled in him the virtue of humanism which he made his own and which continued to be his lodestar throughout his life, as this paper will demonstrate.

From his early years at Leah Street in the East end of Freetown, to his student days at the Sierra Leone Grammar School, early manhood at Fourah Bay College and Oxford, unto his adulthood and prime of his life as lecturer, scholar, literary critic and principal he has always demonstrated the virtues of humanism and love of learning. These are the key traits that have continued to endear him to all who come into contact with him and synthesize to make him into an institution.

LOVE FOR MAN AND COUNTRY

Jones was a humanist even as a child. He was always eager to help the needy and put a smile on the faces of others. We recall how he got flour, sugar, and currants ready to take to Uncle Gordon for the orphans (134). Little did he realize that he will one day become one of the uncles of children's hour.

Jones also used to hold extra mural classes in the city and surrounding villages, for village elders, pensioners, school teachers (66), and appreciated the exhilarating experience of teaching prisoners in Masanke Prison. His choice of 'students' reflects his desire to improve the mindsets of others and encourage them to be the best they can be.

Similarly, the number of adopted children (especially girls) that he and his wife brought up is a measure not only of their altruism but also a reflection of the upbringing of his parents who found it a joy to help other people's children.

His love for mankind is further illuminated in the hosting of some of his friends and their families during the rebel invasion of Freetown. Having opened his doors wide to the families of former colleagues (thirty-seven people in all) fleeing from suspected rebel attack, Jones reveals that he 'turned my study into a schoolroom where we told stories, read poems and made memorable friendships' (85).

During the rebel war, he and his wife had the opportunity either to leave the country or to stay and do nothing. But he chose to stay and make his own input because of his belief that he could not stand 'idly by' if he could do something to change the course of events. He therefore became a member of the Independent Mediation Committee (INAMEC) in order to negotiate with the rebels or to broker peace and help return the country to civilian rule.

REFINEMENT/HUMILITY

One of Jones's endearing traits that a lot of people in high places lack, and a lot more in low places will be the better for, is humility. He is an embodiment of the verse that says, 'Humble yourselves before the Lord, and he will exalt you' (James 4:10, *Revised Standard Version of the Bible*, henceforth *RSV*).

His humility makes him treasure every little thing that others do for him or with him and is forever grateful to his family, friends, teachers, lecturers, colleagues, and many others for the roles that they have played in his life.

This attitude of a grateful and humble heart is reflected in his relationship with the community people that he taught during his extramural classes. He tells us that the community people 'taught me more than I taught them' and referred to his meetings with the prisoners at Masanke Prison as a 'stimulating experience'.

Jones is a warm individual, who makes every experience a thing of joy. He believes that there is always something to be thankful for. A lesson in humility that he learned from his father is worth sharing here:

> He always said to us that there is nothing more rewarding than teaching somebody and finding that he is not only as good as you, but even better. This is a lesson in humility that has guided my relationship with those it has been my privilege to teach. (18)

In days like these when some lecturers and teachers assume they are doing students a favour by teaching them, they should learn from the above that it should be a joy and a privilege not only to teach students but more importantly, to teach those who will surpass them.

Jones is a man of many parts, accomplished intellectually and socially. The way he associated with the great and good at Oxford led to a rich intellectual and social life. There, he took part in bowls, play readings, dramatic performances, sing-song sessions, college sports, cycling, punting, coffee evenings with the chaplain, choir practices, music ensembles, and parties of 'varying shades of grandeur' (50). These social and artistic experiences combine to mould him into a well-rounded and charming individual, comfortably at home anywhere in the world and with all classes of people.

Similarly, he played cricket in Freetown and upon retirement from the field spent several seasons as commentator for international matches. He was also a chorister at Holy Trinity Church, Fourah Bay College, and at Corpus Christi College, Oxford.

To be sure, the life of refinement that he led in Oxford, he replicated in Freetown, where he conducted tutorials in his house '– my wife often producing tea and cakes – played tennis and engaged in extracurricular activities like drama and choral music' (68).

Jones is a widely travelled man with a wealth of experience. He has been to Europe, the United States, Korea, and many African countries, moving from one university, city, country, region, continent to another, giving lectures on his favourite subject, Shakespeare, popularizing the new African Literature, attending seminars, and receiving a lot of awards and honours. His rich life of travels makes one recall Tennyson's 'Ulysses' who delighted in being active:

> Life piled on life
> Were all too little, and of one to me
> Little remains: but every hour is saved
> From that eternal silence, something more,
> A bringer of new things; and vile it were
> ... to store and hoard myself,
> And this gray spirit yearning in desire
> To follow knowledge like a sinking star.
>
> (*Selected Poems*)

Honours and laurels have come to Jones because of his refusal, like Ulysses, to 'store and hoard' himself.

LOVE OF LEARNING

In Sierra Leone the name Eldred Jones is synonymous with education. A man of outstanding intellect, Jones has an enquiring mind and uses all his experiences as a springboard to further improve himself and others. *The Freetown Bond* has revealed that Jones's love for education is not a sort of epiphany. It is in his DNA. We learn from his memoir that

> One of the best memories of my grandmother's cottage is that the walls of the tiniest room were papered with pages from the weekly edition of the London *Daily Mirror* which I read voraciously, sometimes standing on my head to read those pages which had been pasted upside down. (5)

This playful yet singular action to read something in this way not only marks him out as a determined person, but symbolically foreshadows his love of learning that will be associated with his name later in life. His efforts to imbibe knowledge are worthy of emulation in an age like ours where the printed word is fighting for supremacy against the digital world of films and social media. His love of learning is also revealed when he tells us that in his early school days he bought *The New Method English Dictionary* from Mr George Tregson-Roberts (one of his teachers), thereby paving the way to fuel his lifelong addiction to dictionaries, words, writing, scholarship (19).

Jones's life is a tribute to what a man can accomplish with education, foresight, and commitment. His intellectual accomplishment certainly opened doors for him. Achebe's proverb that says 'if a man washes his hands he can eat with kings' (*Things Fall Apart*: 95) is relevant here. All the stories of historic places that his cousin Minnie brought with her from London and what he heard from his pen friend (*The Freetown Bond*: 7) pale in comparison to what he experienced personally, travelling all over the world and meeting with the great and good in society.

Jones tells us that, at Oxford, he was taught by great writers and critics – Clifford Leech, C.S Lewis, Helen Gardner, F.W. Bateson, J.B. Leishman and men of like stamp, and undertook studies on other subjects or books that were not part of his immediate field of study. His lecturers largely shaped his mind both literarily and socially. He makes it quite clear that his studies at their feet, especially F.W.Bateson's, heightened his interest in Shakespeare and eventually led to the writing of *Othello's Countrymen* (1965) and *The Elizabethan Image of Africa* (1971).

Additionally, he loved the Oxford education and viewed it as wonderful experience:

> As one browsed through freely in Blackwell's Oxford's leading bookshop, a book might lead down an alleyway into a new interest … In the carrels of my own college library, I spent many hours with the volumes of the Oxford English Dictionary following the journeys of individual words down the meanderings of English History. It was this freedom and still is, I hope, that helped to make an Oxford education such a liberating experience. (54)

The above quote captures and underpins in a very graphic way Jones love of learning. His excitement is contagious. One can almost picture him chasing the words all over the place until he squeezes their meanings and history out of them. Even the length of time that he spends with the dictionary is truly amazing. By the same token, the publication of *The Krio-English Dictionary*, with Clifford Fyle the Father of the Sierra Leone National Anthem, underscores Jones's interest in the Krio language and his passion for education and research.

Such is his love for learning that even when his vision started to deteriorate, he was more concerned with missing his lectures than with his eyesight. It is also important to look at his love of learning from another angle. I noted earlier that he learns from all his experiences. His feeling of empathy enabled him to admire a few individuals who had lost their vision, and thus subconsciously learned to overcome his 'impediment with fortitude and a positive response'. (162)

JONES THE CRITIC

Jones's humanist outlook is woven into all his activities. Evidence from *The Freetown Bond* and his other writings demonstrates that he is a thorough scholar of a very high order who, through sheer hard work, left an indelible footprint in all the countries that he visited. He is the John the Baptist (Matthew 3, *RSV*) who took it upon himself to pave the way for the flowering of African literature, by founding *African Literature Today* and teaching the works of this new breed of writers in many countries.

It is a tribute to his calling as an educationist that he founded the prestigious journal, *African Literature Today (ALT)* in 1968 and edited it for almost 32 years with his wife Marjorie and Prof. Eustace Palmer, producing very high quality work and helping to promote African literature and African writers. Before he finally passed the mantle over to Prof. Palmer, he edited the journal conscientiously and wrote editorials that wove the various strands of the articles and reviews in each journal into a beautiful tapestry that shed light on the issues dealt with in the journal. It is important to point out that as a humanist critic he not only helped to showcase the writings of the new African breed of writers, but also supported budding writers to publish their articles. This is reflected in *ALT 1*, where Jones underscores the purpose of the journal thus: 'to provide a forum for the examination of African literature, to open the literature to both academic and general readers particularly within Africa itself' (140).

By the same token, as a critic and humanist, his desire to promote others is reflected in the way he made room for writers who could not adhere to the strict formats during the period in which he edited the journal. Jones, as usual made room for others to shine.

In fact, some editions are purposely written as a sort of outdooring ceremony that introduces writers to the outside world and encourage readers to pay more attention to them. Note Jones's assertion in one of his editorials:

> This number is largely devoted to the work of writers whose works have not received much critical attention either because they are relatively new, or because not being what might be described as mainstream, they may unintentionally, perhaps, have been damned with faint praise or neglect. (*ALT 12*)

This is the voice of someone who has *homo sapiens* at heart, someone who yearns for a better world for all and is determined that others should be given a chance to flourish.

In a similar vein, one wonders what might have happened to some of Soyinka's complex works, had Jones not created the platform for them in his lectures, journal and *The Writing of Wole Soyinka*, an invaluable piece of work that sheds light on them. Jones's efforts (and others have joined him) to unveil the beauty and illuminate the importance of Soyinka's works have paid dividend. Soyinka's name resounds far and near and was the first African to be awarded the Nobel Prize for Literature.

Additionally, Jones's dedication to scholarship led him to visit Nigeria and Kenya in order to 'organize a series of visits to art centres, and to talk to as many sculptors, painters, weavers, dyers and writers as possible' (110). Further, he visited 'Obas in their palaces, witnessed Shango and other festivals, and experienced almost at first hand, the traditions from which the modern practitioners derive their inspiration' (112).

This can be viewed as a sort of immersion technique, whereby Jones immerses himself into the culture, arts and social activities of the

Nigerians and Kenyans, in order to understand where their writers come from, and to write his scholarly works from a position of authority.

Additionally, Jones broke new ground when he published *Othello's Countrymen* and *The Elizabethan Image of Africa*. He was not only the first African to have written about blacks in Shakespeare, he was also apparently the first person in the world to have done so. It appears that some whites victimized Othello and left him hanging in the pages of the tragedy as soon as Shakespeare immortalized him. Othello had to wait for close to four centuries before he is rescued and allowed to come to the limelight by Jones, a Sierra Leonean. In the two critical works mentioned above, Jones as it were, broke into a sort of forbidden kingdom that the whites had preserved for themselves. The tribute paid to him by the public orator on the occasion of being honoured jointly with his wife with the Distinguished Africanist Award is relevant here:

> Eldred Durosimi Jones was the first person ever to write about Africans in Shakespeare and single handed he has sparked a research field which he still dominates. In 1965, not long after his country's independence, his seminal work characterizing an African view of Shakespeare's Othello called *Othello's Countrymen*, a study in English renaissance drama appeared. It was a timely and for then a controversial and certainly innovative contribution to Shakespearean scholarship, and other works in this field appeared, including in 1979, *The Elizabethan Image of Africa*. Shakespeare was now finally freed from being the exclusive property of the white Anglo-Saxon world. (*The Freetown Bond*: 64–65)

By the same token, Jones's name is always linked to worthy causes. It is in this spirit that he took up the challenge of chairing the Noma Award for Publishing in Africa when the opportunity presented itself (146).

By allying himself with this cause, Jones as usual helped to give visibility to other African writers who would otherwise have remained unpublished (and invisible), thus breaking the publishing barrier that existed. This is the sort of gap that Prof. Osman Sankoh, Publisher, Sierra Leone Writers Series (SLWS), and Mr Mohamed Sheriff, Pampana Communications Publishing, to name but two, have filled in present-day Sierra Leone. Both Prof. Sankoh and Mr Sheriff have responded to the cry of the Sierra Leone writers (especially the Sierra Leonean university lecturers who need publications for upward mobility) for local publishing houses and have set up publishing houses in which they issue work after work of established as well as budding writers, thus making a very significant contribution to the literary landscape and the development of the nation as a whole.

Jones remains a great figure in Sierra Leone and the international community. He deserves to be celebrated for his life's work and all that he symbolizes. He will always be remembered for his sharp mind, immense contribution to scholarship, and his humanity. To his family and dear friends that know him outside the walls of academia, he will be remembered as a dynamic and witty gentleman who takes everything in his stride notwithstanding his blindness, the loss of his dear wife Marjorie (6 September 2015), the burning of his house and especially his library (20 January 2017). His quiet courage in the face of adversity is captured in the way he describes his loss of vision as a 'great inconvenience' and 'by no means a near death experience' (162). Such fortitude, grace and calmness, perhaps like that exhibited by Uchendu in *Things Fall Apart* are largely responsible for his longevity.

He remains an inspiring figure and a son that Sierra Leone is proud to celebrate. An iconic individual with a sharp mind and a big heart, Prof. Jones is an institution of refinement, wit, reasonableness, humility and fairness, whose legacy spreads far beyond the borders of Sierra Leone.

The question raised by Nathanael in the New Testament regarding Jesus is relevant here: 'Can any good thing come from Nazareth?' (John 1:46 *RSV*). Some may have asked a similar question about our small beloved country Sierra Leone. As the life of Jones demonstrates, something good can come out of Sierra Leone. Something good has come out of Sierra Leone.

WORKS CITED

Achebe, Chinua. *Things Fall Apart*. Oxford: Heinemann Educational, 1958: 95.

Grayling, A.C. 'The Milk of Humanist Kindness'. *The Guardian*, 21 November 2006.

Jones, Eldred D. (ed.). *African Literature Today 1*. London: Heinemann, 1967.

—— (ed.). *African Literature Today 12*. London: Heinemann, 1982.

—— *The Freetown Bond: A Life under Two Flags*. Woodbridge: James Currey, 2012.

—— *The Writing of Wole Soyinka*. London: Heinemann, 1973.

—— *Othello's Countrymen*. London: Oxford University Press, 1975.

—— *The Elizabethan Image of Africa*. Charlottesville: University Press of Virginia, 1971.

Revised Standard Version of the Bible (RSV). The British and Foreign Bible Society, Great Britain.

Tennyson, Alfred Lord. *Selected Poems*. London: Penguin, 1991.

Chukwuemeka Ike

An Administrator with a Cinematic Imagination

AUSTINE AMANZE AKPUDA

Born Vincent Chukwuemeka Ike (28 April 1931 – 9 January 2020) and descended from Ikelionwu, the founder of Ndikelionwu in Anambra State, Ike numbers among Nigeria's most misunderstood and, from the perspective of literary scholarship, under-appreciated novelists. For instance, Chinua Achebe asserted during the public presentation of Chukwuemeka Ike's *Expo' 77* (1980) that 'Chukwuemeka Ike has explored a wider spectrum of Nigerian life than any other novelist. Each of his ... novels takes a different perspective of our story, drawing its material from a different facet of his own rich and diverse experience' (quoted in Ezenwa-Ohaeto *Chinua Achebe: A Biography*: 219), Ernest Emenyonu would lament sometime in 1991 that 'for too long, Chukwuemeka Ike has remained untouched by African and indeed Nigerian critics' ('Of Visions, Youth and Future' 118).

There is no doubt that the recognition of what Achebe identifies as Ike's 'rich and diverse experience', which has rubbed off on virtually all his eleven novels, influenced Ezenwa-Ohaeto's description of Ike as 'the relevant Novelist'.[1] A closer interaction with Ike's novels provides a mosaic that brings up the validity of David Cecil's postulation when writing about Victorian novels that 'a good novel is not just a work of art, it is also a picture of life'. Without a contextualization of Chukwuemeka Ike's writings within the body of Nigerian literature, it will be difficult to appreciate why the evocation of an atmosphere of palpable verisimilitude in his novels has endeared him very much to the general reader and the few scholars who have engaged the near-Dickensian flavour of his canvas.

Whatever backgrounds that have impacted on the realism portrayed in his works seem to derive from the acuteness of the exposure he had in his formative years, especially while receiving primary and secondary education at the Aro Settlement Primary School, Ndikelionwu, Anambra State and the C.M.S. Central School, Ife, Ezinihitte in

Mbaise, Imo State, among others. Experiences encountered at these schools and the Government College, Umuahia must have made his artistic consciousness to register optimally. As he reveals in the course of an interview with B.E.C. Oguzie,

> the world in which I grew up featured in *Toads for Supper*, *The Potter's Wheel*, *The Naked Gods*, and *The Bottled Leopard.* Scenes from Ndikelionwu had featured in the other novels, but under fictitious names … my father had sent me to live with a cousin. The cousin was very different from Teacher in *The Potter's Wheel,* but the years I lived with my cousin at C.M.S. Central School Ife-Ezinihitte and C.M.S. Central School, Nnewi gave me an insight into the life of a houseboy in the early 1940s. ('An Interview with Chukwuemeka Ike': 366)

Chukwuemeka Ike's phenomenal appearances as a member of the editorial team of the famous *Government College Umuahia Magazine* in the 1948/49 session and as co-editor with Christopher Okigbo in 1949/50 of *The Athena*, their School House magazine are indications that his attendance of supposedly rural schools at Ndikelionwu; Ife Ezinihitte, Mbaise; and Nnewi was not a liability. Ike's short story, 'In Dreamland' written when he was 17 years and published under the name V.C.N. Ike in the second volume of the *Government College Umuahia Magazine* is remarkable not only in its adoption of a Shakespeare quote from *A Midsummer Night's Dream* as epigraph but also in its display of intimations of the young writer's familiarity with the dynamics of the modern short story. It is quite strange that despite his privileged exposure to and study of a lot of Juvenilia by Nigerian writers during their high school days, Bernth Lindfors' eagle eyes fails to capture and study in his *Early Nigerian Literature* Ike's first published short story.

However, Lindfors' familiarity with some of Ike's writings while at the University College, Ibadan is intriguing, Concerning Ike's 'Nationalism versus Religion' published in *Bug*, Vol. 4 No. 4 of 24 January 1953, Lindfors remarks that Vincent Chukwuemeka Ike was 'responding with vigor to a sermon by a European member of Staff' (83). What could be discerned as the relatively subdued tone of the disposition of the 'strong man of the pen', noticeable in Ike's 'In Dreamland' assumes a greater propensity in his University College, Ibadan essay. Lindfors reports how Ike tasted what was his first experience of censorship. According to Lindfors, 'attacks by V.C. Ike and others on a sermon given by a European member of staff in 1952 had led to the banning of the *Bug* for four months in the spring of 1953' ('Popular Literature for an African Elite': 86). It is probably this type of suppression of the youth that Chukwuemeka Ike campaigned against not only as an administrative staff at the University College, Ibadan but also at the University of Nigeria, Nsukka where he rose from being a Deputy Registrar to becoming the first African Registrar. As a novelist, Ike continues this protest not only in *Toads for Supper* but also in *Our Children Are Coming!*

With roughly six published short stories between 1948 and 1963 before his first published novel, there are indications that the young writer was test-running his passion through the medium of the short story. That his first attempt at mastering the form of the short story as seen in 'In Dreamland' turned out successful may perhaps have drawn from the confidence he had found after scoring '27 out of 30' in an essay focused on the 'Eclipse of the sun' (see Oguzie: 367). It is salutary that, in evaluating Ike's relatively accomplished juvenalia, 'In Dreamland', Terri Ochiagha points to how the short story embodies a 'dream motif [that] is a hallmark of Ike's adult writing; *The Bottled Leopard*, *The Naked Gods*, *Conspiracy of Silence*, and *The Potter's Wheel*, all [of which] reveal a similar interface between the conscious and the unconscious' (*Achebe and Friends at Umuahia*: 119). With regard to the early promise demonstrated in 'In Dreamland', it is a major loss to Nigeria's literary history that there is no bound copy of Ike's published and uncollected short stories.

The tendency by readers of Chukwuemeka Ike's novels to identify and match real-life characters with the novelist's fictional creations, no doubt, constitutes the greatest tribute to Ike's three-dimensional characters simulation. Invariably, the unique writing style that makes readers insist on a particular flesh-and-blood character, as the model that he sets out to portray fictionally qualifies Chukwuemeka Ike as probably the most seemingly biographical or factional novelist in Nigeria. His profile as an accomplished artist whose fictional portraits appear palpably realistic could be seen as the converse of the story told about Soren Kierkegaard who led a life that appeared to resonate with the same dramatic intensity as that of the fictional Prince of Denmark in William Shakespeare's *Hamlet*. The creation of memorable characters, who oftentimes remind readers of real-life persons they are familiar with, is an experience that many readers have associated with Ike's novels, especially *Toads for Supper* (1965), *The Naked Gods* (1970), *The Potter's Wheel* (1973), *The Chicken Chasers* (1980), and *The Bottled Leopard* (1985). For instance, as a result of the Dickensian brush with which Ike paints Mrs Ikin in *The Naked Gods*, it is reported that, in real life, the husband of the woman so fictionalized as Mrs Ikin had to separate from her. According to Ike's reflections, 'Encounters with Readers and Critics', long after 'the first Professor of Economics at the University, Professor N.A. Cox-George, returned to his home country, Sierra Leone, in the mid-1960s', *The Naked Gods* (1970), which people consider a satire on the University of Nigeria, Nsukka was published. As Ike relays concerning a very touching matter,

> when *The Naked Gods* came out in 1970, some people who knew the Cox-Georges at Nsukka identified Mrs Cox-George with Mrs Ikin, a character in the novel. On a visit to Sierra Leone in the mid-70s, I felt awkward travelling with Professor Cox-George in the shuttle from Lungi Airport to Freetown because I did not know whether he had read *The Naked Gods* and whether he subscribed to the view that I had portrayed his wife as Mrs. Ikin. (327)

Arising from such unintended impact of Ike's satiric pen, Ike reveals that 'during the ferry portion of the journey, he [Professor Cox-George] walked over to where I sat and advised the gentlemen chatting with me to watch what they said in my presence or they might feature in my next novel!' (327)

It will be difficult to discuss the concept of folklore in Nigerian literature or perhaps metaphysics in the same literature without including a reference to Chukwuemeka Ike's *The Bottled Leopard.* A striking development here is that although the novelist is focusing on an aspect of the Igbo worldview that accommodates the mystical experience of leopard possession, it is an experience often associated with and limited to some communities in the present Akwa Ibom State. In the aftermath of the mysterious leopard killings of the 1940s within the designated 'leopard area' and which provoked the massive campaign for the destruction of leopards in the late 1940s (see Nwaka 'The "Leopard" Killings of Southern Annang, Nigeria, 1943–48': 434) to discourage the leopard cults, especially among the Annang, a precocious student at the Government College, Umuahia could have decided to investigate the phenomenon among the Igbo which had communities that believed in the same metaphysical experience. For instance, not many would know that while he was residing in Uzuakoli in the late 1940s, Pita Nwana killed one of the many leopards, traceable to his expertise as a distinguished hunter; efforts were made to ensure that there was no missing 'leopard man' before the author of *Omenuko* was celebrated as a leopard slayer. Whatever led Chukwuemeka Ike to consult some leading members of the leopard men cult must have equally brought about his domestication in *The Bottled Leopard* of the Igbo heritage to which he was privy. However, rather than employ a fictional character who is an old man like Obinkaram Echewa's old man, Ahamba, who deliberates on mystical and spiritual experience with Father Higler in *The Land's Lord,* Ike prefers to guide his readers through the eye of an informed adolescent.

The supposed remoteness of the experience of having a man incarnate a leopard is what Amobi Ugochukwu presents to readers of Chukwuemeka Ike's *The Bottled Leopard.* And it is instructive that Amobi compares the Judeo-Christian tradition with the Igbo culture with which the novelist was familiar. Although he failed to acknowledge whatever interactions he might have had with a sister, Adaeze Ulanwa Ike, who was a 'leopard woman' and mother to Peter Nwankwo and Professor Chimalum Nwankwo, Chukwuemeka Ike confesses to Ezenwa-Ohaeto that, considering 'my interaction with Dibia Agu [which] provided me with an authentic physical setting for part of the novel' (88), there is no doubt that 'I used fact to build fiction' (89).

Beyond the inclination to demonstrate how some children in *The Potter's Wheel* and *The Bottled Leopard* share the same 'healthy appetite for mischief' that we associate with George Eliot's Eppie in *Silas Marner*, Chukwuemeka Ike creates the space for his child-characters to reflect on such mystical topics as wizards, witches, and possession. However, without being trained as a theologian, Amobi demonstrates an awareness of the correlation between 'the story of a man possessed by a demon, and how Jesus Christ transferred the evil spirits from him into swine' and the phenomenon of leopard possession (*The Bottled Leopard*: 52). As Ike's narrator presents Amobi's mindset, 'And each time, he had seen a relationship between the story and the claims at home about men possessing leopards' (52), Ike artistically guides the highly inquisitive Amobi to query their Teacher nicknamed Iambuc.

Amobi discountenances the assumption that he must have been 'asking about wizards and witches' (53) as alluded by Karibo. We see a highly articulate and precocious Amobi raise questions 'about the New Testament lesson read in the service yesterday sir' without considering the Teacher's 'rule on question time' (52). Despite all the efforts made by Iambuc to restrain Amobi, the boy insists on finding the right answer to his own question(s):

> Is it possible, sir ... for a man's spirit to go into an animal temporarily and control the behaviour of that animal? ... my question is whether it is possible for a man to send his spirit into an animal, like a leopard, to control that animal for some time. (52)

Only a knowledgeable novelist with the insights of a child psychologist can recreate the memorable characters we have in *The Bottled Leopard.* It is a trait that goes back to *Toads for Supper* and *The Potter's Wheel*, and one which makes Ike an expert in writing the Bildungsroman sub-genre.

Chukwuemeka Ike's *Sunset at Dawn* (1976) is one of the Nigerian war novels that enjoys the rating of scholars and ordinary readers. In this novel, Ike satirizes the unusual appropriation of reluctant and unenthusiastic civilians into the army and who not only make a caricature of themselves through spotting the Fidel Castro beard and Biafran suits but also engage in less than worthwhile assignments. The subject of the inter-ethnic marriage between Dr Amilo Kanu and Fatima is a sub-plot that perhaps would not only influence an aspect of Chimamanda Adichie's drama text, *For Love of Biafra* but also her war novel, *Half of a Yellow Sun.* It is possible that Adichie's indebtedness to Ike's war novel, especially in the sense of the love tango between Dr Kanu, an Igbo-Biafran and Fatima, an Hausa-Nigerian, gave rise to the creation of Adaobi and Mohammed in Adichie's *For Love of Biafra.* The Mohammed presented in *For Love of Biafra* is as compassionate as the one presented in *Half of a Yellow Sun.*

Notwithstanding Willfried F. Feuser's assertion that *Sunset at Dawn* does not mark 'the great novel on the conflict' ('Chukwuemeka Ike, *Sunset at Dawn*, A Novel about Biafra': 233), there is no doubt that it is still an important war novel. It is one of the few that produces songs that made the soldiers fighting on the side of Biafra feel like conquerors. We encounter a sustenance of this same tradition of evoking such martial drill songs in Peter Onwudinjo's *Women of Biafra and Other Poems.* Elsewhere, despite insisting that 'the great Nigerian war novel is yet to be written' ('Witness'104), Ernest Emenyonu nonetheless concedes that 'Ike tries to touch on all issues and all forms of experience in Biafra in the 246-page novel. The pace of the narrative is solemn and dignified and the reader's suspense is held till the last page' ('Witness' 100).

Ike's idea of '*Chi iwere ehihe jie*' (literally sunset at dawn) is quite apt and resonant in the several near-real-life documentations of the anti-Biafra lives propagated by the characters in the Biafran enclave. Among the Nigerian war films in circulation, it does appear that Ike's war novel might have impacted on the creation of the inter-ethnic marriage between the two main characters in *The Battle of Love*, Dubem and Habiba, and the evocation of certain aspects of the recruitment of civilians and the Obodo setting of the novel in the film, *Turning Point.*

Chukwuemeka Ike is a celebrated African author of a brand Western literary critics usually refer to as 'blue jeans' literature. He is one of Nigeria's greatest promoters of what could be discerned as a youth agenda in his novels. Although many of Ike's eleven novels present accommodating portraits of childhood and adolescence, some of the most popular ones are *Toads for Supper*, *The Potter's Wheel*, *The Bottled Leopard*, *Our Children Are Coming!* and *Toads for Ever*, the follow-up to his first novel, *Toads for Supper.* It is exciting to know that it is the same Aduke and Amadi story, the logical conclusion of which many of Ike's readers clamoured for in *Toads for Supper* that gave rise to Ike's last novel, *Toads for Ever* (2008). In 'Of Visions, Youth and Future: Chukwuemeka Ike's Recent Writings – *Our Children Are Coming!* and *The Search*', Ernest Emenyonu accounts for Ike's qualification as a memorable writer of youth – oriented literature. According to Emenyonu, Ike's

> wealth of experience in his career as a Registrar in university administration as well as in the West African Examinations Council more than best equips him to write with authenticity on the psychology of youth development. Indeed Ike's most impregnable arena as a novelist is the portrait of the crisis of youth development in the Nigerian society. (119)

The commendation that Emmanuel Ayandele gives Chukwuemeka Ike as the author of the two major university-centred novels, *Toads for Supper* and *The Naked Gods*, is enough evidence that very highly placed Nigerian educationists consider Ike as a significant novelist. Beyond his career profile as a prominent historian and university lecturer, Professor Ayandele was also an established administrator. He was Acting Principal (VC), Jos campus of the University of Ibadan, 1974–75; Acting Vice Chancellor and later Vice Chancellor, University of Calabar, 1975–82. Although he was not a figure in literature in the conventional sense, Professor Ayandele was definitely part of the crème de la crème of the Nigerian elite. Where an Albert Einstein would testify that 'Dostoyevsky has given me more than any scientific thinker does, more than Gauss' (*The Idiot* Book One, jacket sleeve), Emmanuel Ayandele, in contextualizing the basis for his discussion of the educated elite in Nigeria, affirms his indebtedness to Chukwuemeka Ike, 'whose exposure of the other side of the undergraduate in *Toads for Supper* and of the don in *The Naked Gods* is vivid in the minds of many' (1). Elsewhere, in a 23 March 1990 letter, Professor Ayandele, in commending Ike's *The Chicken Chasers* 'which kept my spirits aloft in Conakry', notes that, 'the one idea that occurred to me was the salutary impact its transformation for the television screen could have on Nigerian viewers' (quoted in Ike 'Encounters with Readers and Critics': 333).

What Ayandele says about the possibility of transforming Ike's *The Chicken Chasers* into a television production is an important comment that should open the eyes of readers to the dramatic character of Ike's novels. It is within this context that one should appreciate why Ernest N. Emenyonu's recognition that Chukwuemeka Ike has 'a good ear for dialogues and a penchant for accurate minute details. Often these minute details enrich his dialogues' ('Of Visions, Youth and Future': 123) should not to be taken for granted. Similarly, Emenyonu's assertion that Ike is 'at his best when reporting dialogues' (124) should not be seen as an accidental artistic virtue and development. Rather, such attribute of making his novels read like drama performances is a heritage traceable to Ike's Government College, Umuahia days as a member of the Drama Club.

It is quite regrettable that the Chukwuemeka Ike, whose dialogues in *The Potter's Wheel*, *The Chicken Chasers*, *Sunset at Dawn*, *The Bottled Leopard*, and *The Search*, among others, are quite engaging, did not publish a drama text. Beyond the much advertised affinity he and his friend and dormitory mate, Christopher Okigbo, shared with Mr Charles Low, the affable theatre-oriented Australian teacher at the Government College, Umuahia, what Charles Low did with the Drama Club during Ike's studentship is a heritage that dates back to the 1930s when vibrant drama seasons were part of the regular fare at Umuahia. As such, if in some of his novels Ike could simulate dialogues that actually read like they were scenes from a drama presentation, it is mainly because of the exposure that he had at his alma mater. After all, according to Ike's testimony in the interview he granted B.E.C. Oguzie, 'At Umuahia also, we had teachers (or masters as we called them). One of them, an Australian, Charles Low, who was a poet and playwright, got my class to work with him on a play, *White Flows the Latex, Ho!*' (367). It is instructive that Barth Oshionebo has documented how, long before Ike's generation at the famous college, their predecessors had mounted very inspiring theatrical productions that appeared to be out of the world. Among the remarkable productions by the students of the Government College, Umuahia, according to Oshionebo, are the scenery for *Noah's Ark* in 1936 and in 1939, the stairways in *Joseph and His Brethren* (15). No doubt, the verisimilitude generated through the stage designs is a complement to the language realism created not only in the dialogues of these plays but also the ones Chukwuemeka Ike witnessed or participated in.[2]

It is instructive that Mcpherson N. Azuike, one of the most informed and accomplished stylisticians in Nigeria, takes a special note of 'the dramatic effect and atmosphere achieved through the use of dialogue' in Ike's *The Bottled Leopard* ('Language and Style in *The Bottled Leopard*': 61). Azuike considers conversations involving Dibia Ofia, Amobi, and Amobi's father, as seen on pages 127 and 128 of the University Press edition he read as, remarkable. Little wonder then, that after examining these encounters, Azuike affirms that 'this excerpt reminds one of theatrical language. In it the participants evoke an atmosphere of lively encounter and palpably dramatic effect' (62).

Despite being born to what was considered an autochthonous ruling family, Chukwuemeka Ike did not display tendencies towards aristocratic behaviour. It is possible to trace his seemingly pacifist approach to traditional leadership to certain designs and portraits in his novels. Although there are indications that Chukwuemeka Ike never threw up his profile as a traditional ruler to intimidate anyone, I was taken aback at the extremely formal and near-ritualistic manner Osita Ike instructed me to address his father when I sought a telephone interview with him. Not given to the type of tedious greeting format Osita insisted that I should observe, I almost cancelled the interview. However, when I engaged the famous novelist and administrator, there was no attempt at editing what I could have got right or wrong by way of being courteous. Chukwuemeka Ike displayed the razor-sharp mind he had nurtured from the time he lived with 'Teacher' at the C.M.S. Central School, Ife and beyond by reminding me that my name was familiar and that he had read the book I edited on Charles E. Nnolim, *Reconstructing the Canon: Festschrift in Honour of Professor Charles E. Nnolim* (2001). The telephone interview sessions we had had offered him the needed opportunity to address the unfinished dialogue on his own novels that I had had with Professor Nnolim. His reservations about Nnolim's refusal to respond to my lead questions on Ike and the Nigerian canon and the disposition that could have given rise to Ike's inimitable rejoinder to Bernth Lindfors, 'The Canon according to Bernth Lindfors' published in Nigeria's *Daily Times* of 20 June 1992 was also given vent during the incisive interview.

It is the Chukwuemeka Ike who had enough exposure and knowledge as a writer and later, as a Professor, that authored *How to*

Become a Published Writer (1991). Familiarity with the manuscripts of fledgling writers must have impressed on Ike the urgency to do the writers' guide he published in 1991. In the aftermath of a Visiting Professorship of English and Creative Writing at the University of Jos between 1983 and 1985, Chukwuemeka Ike showed the quality he was made of by his editorial interventions on manuscripts of aspiring writers. Here, it is important to reveal that Peter Nwankwo had had the privilege of having Chukwuemeka Ike read through the manuscript of his first novel, *Devil's Playground*, before it was published by Longman in 1989. Peter Nwankwo disclosed to this writer that, despite the extremely busy schedule that Chukwuemeka Ike kept, he, unlike others who were given the same manuscript, wrote a very professional seven-page critique that helped to re-direct the structure of Nwankwo's novel. There is little wonder then that Chukwuemeka Ike would later be professionally involved in programmes aimed at encouraging and nurturing aspiring writers. This is part of the background and philosophy behind his identification with and support for book foundations such as the one he established in Awka.

Chukwuemeka Ike's distinction as a novelist has been highlighted by important scholars such as Ernest N. Emenyonu, B.E.C. Oguzie, Kanchana Ugbabe, Ezenwa-Ohaeto, Mcpherson Azuike, Victor O. Aire, Clement A. Okafor, Godwin Ioratim-Uba, and Obi Iwuanyanwu, among others. However, Terri Ochiagha's evaluation of Ike's significance as a novelist is quite captivating. According to her,

> Ike's thematic range remains nonpareil in Nigerian literature. He has experimented with a variety of novelistic genres, including the school story in *The Bottled Leopard*, detective fiction in *The Naked Gods* and *Expo' 77*, and the travelogue in *To My Husband from Iowa*. He is also one of the most prolific writers: his published output includes eleven novels, several short stories, a fictional travelogue and several books of nonfiction. In 2008, he published *Toads For Ever*, the sequel to his first novel. He is the sole first-generation writer to have translated one of his early novels, *The Potter's Wheel*, into Igbo. This translation, *Anu ebu Nwa*, was published in 1999. (*Achebe and Friends at Umuahia*: 178)

From the foregoing, there is no doubt that Chukwuemeka Ike ranks as one of the most important novelists of Nigerian extraction produced. He is, without any argument, one of the few to stretch the canvas of the Nigerian novel to accommodate more spectacles, scenes and pictures, episodes, memorable events, themes and styles. As such, any discussion of the Nigerian novel that does not grant appropriate and deserving space to Chukwuemeka Ike's fiction whose cinematic imagination resonates many years after an encounter with his fictional works, will be an under-represented appraisal.

NOTES

1 This expression is part of the title of the special lecture Ezenwa-Ohaeto presented to mark Chukwuemeka Ike's 60th birthday. First delivered on the occasion of the 26 April 1991 ceremony to mark the event, Ezenwa-Ohaeto's paper was eventually published in Nigeria's *Daily Times Review of Ideas and the Arts* 4 and 18 May 1991 and, later in, *Chukwuemeka Ike: A Critical Reader* edited by Kanchana Ugbabe in 2001. Without doubt, as one of the influential authorities on Chukwuemeka Ike's novels, Ezenwa-Ohaeto was eminently qualified to describe Ike as 'the relevant novelist'. Ezenwa-Ohaeto's essays (by 1991) on Chukwuemeka Ike included 'The Bottled Leopard', 'Chukwuemeka Ike' (in *Perspectives on Nigerian Literature*), 'Chukwuemeka Ike' (*Daily Times*), 'The Historical Dimension of Chukwuemeka Ike's Female Characters', 'Our Children are Coming', 'Chukwuemeka Ike at Sixty', 'Ike's New Novel of Ideas'. Ezenwa-Ohaeto's stature as a prominent critic of Ike's novels can be demonstrated in the anthologization of his four essays in the eighteen-chapter book edited by Kanchana Ugbabe.

2 As can be seen in Bernth Lindfors' 'Nigerian High School Plays 1950–1972', the privileged exposure to theatre experience which Chukwuemeka Ike and his generation enjoyed at the Government College, Umuahia is a reality that was sustained decades after. One beneficiary of this experience and an acclaimed writer, Dr. A.N. Chukuezi, confirmed this in an interview he granted this writer (Akpuda 'Interview with Dr. A.N. Chukuezi, Father of Modern Drama in Igbo Written Literature'). According to Chukuezi, there is no doubt that Prof Bernth Lindfors' observation of the tradition of regular drama presentations at Umuahia is true in absolute terms. In fact, I was privileged to see a photograph in my late uncle's house (also an Umuahian) of the cast of *The Mikado* an opera by Gilbert and Sullivan, set in Japan they acted in 1952 in which he played a prominent part. So prior to my time at the Government College, Umuahia, dramatic art flourished very well despite the fact that it was not taught as a subject. It is rather an irony that despite all these exposures the school has not produced many playwrights (146–147).

WORKS CITED

Adichie, Chimamanda Ngozi. *For Love of Biafra.* Oxford: Spectrum, 1998.

—— *Half of a Yellow Sun.* Lagos: Farafina, 2006.

Akpuda, Austine Amanze. 'Interview with Dr. A.N. Chukuezi, Father of Modern Drama in Igbo Written Literature', *Enyimba Journal of the Humanities and the Social Sciences* Vol. 1, No. 1, 2007: 145–158.

Ayandele, E.A. *The Educated Elite in the Nigerian Society.* Ibadan: University Press, 1974.

Azuike, Mcpherson N. 'Language and Style in *The Bottled Leopard*'. In *Chukwuemeka Ike: a critical reader*, Kanchana Ugbabe (ed.). Ikeja: Malthouse, 2001: 50–63.

Dostoyevsky, Fyodor. *The Idiot.* Book One. Moscow: Raduga Publishers, 1971 [1868–69].

Emenyonu, Ernest N. 'Of Visions, Youth and Future: Chukwuemeka Ike's Recent Writings – *Our Children Are Coming!* and *The Search*'. In *Studies on the Nigerian Novel*, Ernest N. Emenyonu. Ibadan: Heinemann Educational Books, 1991: 117–132. Repeat in *Chukwuemeka Ike: A Critical Reader*, Kanchana Ugbabe (ed,). Lagos and Oxford: Malthouse Press, 2001: 111–123.

—— 'The Nigerian Civil War and the Nigerian Novel: The Writer as Historical Witness'. In *Studies on the Nigerian Novel*, Ernest N. Emenyonu. Ibadan: Heinemann Educational Books, 1991: 89–105.

Ezenwa-Ohaeto. *Chinua Achebe: A Biography.* Oxford: James Currey and Bloomington: Indiana University Press, 1997.

—— 'Chukwuemeka Ike'. *Winging Words: Interviews with Nigerian Writers and Critics.* Ibadan: Kraft, 2003: 83–92.

—— 'Chukwuemeka Ike at Sixty, Photographic Imagination: The Significance and Celebration of Chukwuemeka Ike as a Writer'. *ALA Bulletin* Vol. 17, No. 4, 1991: 45–48.

—— 'Chukwuemeka Ike'. *Perspectives on Nigerian Literature: 1700 to the present*, Vol. 2. Yemi Ogunbiyi (ed.). Lagos: Guardian Books Nigeria, 1988: 142–146.

—— 'Chukwuemeka Ike: The Relevant Novelist'. *Daily Times Review of Ideas and the Arts*, 4 May 1991: 20–21; and 18 May 1991: 13; and later in *Chukwuemeka Ike: A Critical Reader*, Kanchana Ugbabe (ed.). Ikeja: Malthouse, 2001.

—— 'Ike's new novel of ideas'. *Daily Times*, 29 November 1991: 18.

—— 'Our Children Are Coming'. *Society* Vol. 2, No. 10, 1990: 22.

—— 'Our Children: The Fictive Innovation of Ike'. *Daily Times*, 8 March 1991: 10; and 20 March 1991: 16.

—— 'The Historical Dimension of Chukwuemeka Ike's Female Characters'. *A Current Bibliography on African Affairs* Vol. 22, No. 2, 1990: 137–145.

Feuser, Willfried F. 'Chukwuemeka Ike, *Sunset at Dawn*, A Novel about Biafra'. *African Literature Today 13: Recent Trends in the Novel*, Eldred Jones and

Eustace Palmer (eds). London: Heinemann, 1983: 229–233.

Ike, Chukwuemeka 'Encounters with Readers and Critics'. In *Goatskin Bags and Wisdom: New Perspectives on African Literature*, Ernest N. Emenyonu (ed.). Trenton, NJ: Africa World Press, 2000: 323–346.

—— *The Bottled Leopard*, 3rd edn. Ibadan: University Press, 2004.

Lindfors, Bernth. 'Nigerian High School Plays 1950–1972'. *Kiabara: Journal of the Humanities* Vol. 3, No. 1. 1980: 47–88.

—— 'Popular Literature for an African Elite'. In *Early Nigerian Literature*, Bernth Lindfors. Ibadan: Caltop Publications, 1982: 75–90.

Nwaka, Geoffrey I. 'The "Leopard" Killings of Southern Annang, Nigeria, 1943–48' *Africa: Journal of the International African Institute* Vol. 56, No. 4, 1986: 417–440.

Ochiagha, Terri. *Achebe and Friends at Umuahia: The Making of a Literary Elite.* Woodbridge: James Currey, 2015.

Oguzie, B.E.C. 'An Interview with Chukwuemeka Ike'. In *Goatskin Bags and Wisdom: New Critical Perspectives on African Literature*, Ernest Emenyonu (ed.). Trenton, NJ: Africa World Press, 2000: 365–374.

Oshionebo, Barth. 'Theatrical Design in Nigerian Schools, 1935–1960' *Theater Experience: A Journal of Contemporary Theatre Practice* Vol. 3, No. 1, 2006: 10–18.

Reviews

Sakui Malakpa. *Black Professor, White University.*

Mill City Press: Maitland, Florida, 2020, 285 pp. $16.99
ISBN 9781631291616, paperback

Sakui W.G. Malakpa's book, *Black Professor, White University*, feels fortuitous in the current scene of the American university consumed by its politics of diversity and representation. Darrell Thomas, a Southern African-American, and his wife Vanessa, embark on a very dark journey towards an academic career in a predominantly White University. The story is almost cliché, and so it might seem indeed its outcome, because it is very familiar to 'faculty of color' whose experiences in these American institutions is the general leaven of Malakpa's story. At first glance, just with the title, the book might seem like a memoir; a sort of witnessing into the transactional relationship between the African or Black academic struggling towards tenure and a fulfilling intellectual career in one of what might just be the last bastions of institutional racism in America, the Academy. But Malakpa chose the mode of fiction – and it is realistic fiction; nothing allegorical or symbolic. This choice is interesting in itself because at the very core of Malakpa's strategy is the need to mask, or direct attention to crucial events and real people by fictionalizing the events. Darrell Thomas by this account therefore is a highly qualified, very accomplished young African-American in this story. By the time we meet him in the first pages, he is returning, after a tumultuous academic career to his beloved hometown of Ghetahzia – a fictional 'small, hot, Southern town' – where he was at least highly regarded. He is returning to familiar, and consoling roots; to the embrace of his last surviving siblings, and the deep memories of his parents, Clara and Dwayne who lived their lives out in Ghetahzia and are buried there. In this fictional town, whose name has been taken by the author from his own Loma language to mean, 'we have endured a lot', Darrel Thomas is seeking a form of respite after his many years of 'rocky' endeavour at a Midwestern University also fictionally called SAU – the Southwest Achval University. As the story goes:

> Indeed, there was a plethora of reasons why Darrel wanted to spend some time in Ghetahzia before venturing anywhere else. He wanted to put most of his university experiences out of his memory and there was no better place for that than Ghetahzia.

This assertion basically set the tone of this book, and gives it up very early as an account of this one man fighting for his place in a racially charged, and difficult environment, and has possibly been worsted, and is exhausted. Again, the story is really almost too familiar and predictable.

Chapter One, 'The Interview', describes the real challenges that confront under-represented people in the system of White privilege that American university campuses constitute, in that quite herculean bid to gain entry into the cloisters. Darrel Thomas has outstanding qualifications – a good PhD, two refereed articles already published in outstanding journals, co-Principal Investigator on two major federal grants – yet he has numerous rejections from universities to which he had applied for jobs in his field of Economics. One letter alone came that summer, of his many applications, inviting him to an interview. This, his academic mentor Professor Kwame Lumumba, tells him is very normal and not unexpected. There are longstanding barriers to hiring African-Americans and other historically under-represented populations in traditionally 'White universities'. For a Black candidate, interviewing under these conditions of marginality for academic positions in these universities also places them in the orbit of intense scrutiny; like the external gaze that forces Black folk inside the Duboisian 'veil'. For the Black subject, going to an interview in a predominantly White university is 'akin to standing in the light while others watch you from the dark. Naturally, they can see you but you cannot see them, and know not precisely how they look at you.' If the Black face is an inscrutable mask, the White face is the grin of the skull; equally inscrutable but with its mask of polite and deadly irony. This first chapter puts in very stark context the basic character of the crisis, the contradictions, the hypocrisy of the White American University enclave, and the lip service it pays to issues of diversity, and minority inclusion; a cardinal necessity of the American civil rights credo which is the cardinal gain of the 1964 Civil Rights Act. The institutional subversion of this Act is the subject of the criticism of this book. As the old Professor Dwayne Coleman whom he encounters during his interview on the campus of the University tells Darrell Thomas with sardonic resignation,

> when I came here, I ignored all the rubbish – I mean, the horrible, and racially hostile campus atmosphere – and got myself tenured; now I feel; I feel locked into a hellish prison and articulated or not, that's the feeling of many people from under-represented groups otherwise called people of color.

This pretty much sums up the truth of this book. Malakpa's book is possibly an example of the genre 'Faction' – that hybrid mode of narration that unites fictional truth with factual, first-hand experience. A *memoir-noir*.

Chapter Two 'Wow, Starting Here!' – details Darrell Thomas' acceptance of the position at SAU, having also negotiated, as part of his acceptance, a position for his wife as a non-tenure track lecturer in the same University. Moving from Ghetahzia, and looking for a house to settle in, in that small university town seventy miles away, gives credence to another quite predictable story of racial difficulties. The problem of settling to an embrace of new neighbours. There are also some guiding lights into the complexity of human behaviour. The Thomas family is welcomed into their new neighbourhood with flowers and open arms by an old White couple, but ignored by the lone Black family down the road. The hints are crucial. Malakpa attempts to divest his angst of any essentialist, racially narrow constructions, because we see not only in this couple the statement that Whiteness is not the problem, but the institutionalized form of 'white power' and those who defend the privilege based on the idea that accentuates 'Whiteness seen as property'. There is also the White Professor Andy Barclay, who proves in the course of the story to be a bastion of integrity and support, and who defends Darrell Thomas as he confronts the institutional racism and the traps set to destroy his chances of tenure and advancement in the university.

Chapter Three – 'Guidance for Settling in' – essentially provides a *tour-de-force* of Black intellectual strivings, triumphs and sacrifices within the context of America's segregationist history; from Alexander Twilight, the first African-American in the United States to graduate from College in 1823 from the Middlebury College in Vermont; he was also the first African-American to serve in public office, elected to the Vermont Legislature; there was Richard Theodore Greener who in 1870 became the first African-American to graduate from Harvard; in 1876 Edward Alexander Bouchet became the first African-American to earn a PhD, earning his doctorate in Physics from Yale; W.E.B. Du Bois became the first African-American to earn a PhD from Harvard in 1895. In 1849, Charles L. Reason was the first African-American to become a College Professor, while Bishop Daniel Payne became the first African-American College president at Wilberforce College, Ohio, in 1856. Also Sarah Jane Woodson Early was the first African-American woman College Professor, while Mary Jane Patterson, became the first African-American woman to earn a Bachelor's degree, graduating from Oberlin College in

1862; in 1921 Sadie Tanner Mossell became the first African woman to earn a PhD; earning her doctorate in Economics from the University of Pennsylvania. In the same light, the first Native American woman, Susan La Flesche Picotte, who graduated from the Women's Medical College of Pennsylvania in 1889, was the first to graduate in medicine. We also learn about Derrick Bell, one of the pioneers of the Critical Race Theory whose *Race, Racism, and American Law*, does in a sense provide a critical ideological interface to the aim of Malakpa's book. As Dean at the Oregon School of Law, Derrick Bell resigned his position in protest against the School's unwillingness to hire an Asian-American woman on the Law Faculty. In a sense, Professor Bell also cleared the grounds for the hiring of Lani Guinier as the first African-American Woman to join the Harvard Law School as a tenured faculty member, following Derrick Bell's resignation from the Harvard Law School in protest of its hiring practices that also excluded African-American women. These were the Black 'vanguards in the professoriate'. As the old Professor Kwame Lumumba, the moral and intellectual guide for Darrel Thomas in this book says to him: 'before you ask "why me," or "how can people be blatantly racists and discriminatory," think what those people went through to achieve in the face of *de jure* racism.'

Chapter Eleven of Malakpa's book amplifies the conceptual universe framed in Chapter Three, by exploring the literature essential to comprehending the status of Blacks and other marginalized groups in the American Academy – from Gloria Bonilla-Santiago's, The *Miracle on Cooper Street: Lessons from the Inner City*, Margo Jefferson's *Negroland*, Damon Tweedy's *Black Man in White Coat: A Doctor's Reflection on Race and Medicine*, Imani Perry's *More Beautiful and More Terrible: The Embrace and Transcendence of Racial Inequality in America*, and Eddie S. Glaude's *Democracy in Black: How Race Still Enslaves the American Soul*, to Ta-Nehisi Coates, *Between the World and Me*.

The rest of the story, up to the last Chapter Twenty-five, gives coverage to the essential nature of the conflict, which underscores Black anguish; the general Black faculty feeling in the US university system, but particularly in the predominantly White university, of being trapped in the 'hell' of White intellectual privilege. We get a vivid sense in the struggles of Darrell Thomas, of the bruising battles to achieve tenure, and claim the fair balance of a level playing field, between Black and other coloured aspirants to the professorate. They fight for the security of their tenure; the dignity of their intellectual labour, against the hypocrisy, discrimination; isolation, underhanded methods; and other institutional traps laid before Black faculty to undermine their careers in these truly 'Ivory' towers.

Although this book is written as fiction, it is not meant to be strictly read as fiction. It does in fact include a bibliography and footnotes. It merely adopts the mode of the novel to convey essential truths. The aim is possibly to hide the writer from the task of naming real subjects in this devastating witnessing to truth. But as fiction, it fails in many spectacular ways. There are many technical infelicities that mar the book. First is its pacing. There are many intrusive asides that clearly signal a rather amateurish handling of the narrative transition between dialogues. It requires a mastery of the craft of narrative to create those credible, realistic pauses, and the transitions that formalize the nature of speech in a story. Sometimes the dialogue is far too plodding; too heavy and staged to be realistic. There is also the chronology: time here is not properly controlled. For example, we see Darrell Thomas, old and retiring at sixty-eight at the beginning of the story. He returns to Ghetahzia at the end of his career; but the chronological shift is problematic, and the action of the story is in a time-frame that is logically incompatible with the narrative frame. This disrupts the organic order of the book. In general, this story could have been better edited, and tightened. Then, again, Sakui Malakpa has rendered far more a sociological account of a phenomenon within the academy that has, since the Black Lives Matter Movement in the United States, gained new importance and vital immediacy: the status of Blacks in public institutions long controlled by White folk. The question of diversity, inclusion, and equity; the power of White gate-keeping in powerful institutions which have long contained, isolated, limited, and marginalized under-represented groups in these institutions of long-tenured White privilege, is a matter of unending challenge for Africans and their descendants in America. The politics and practice of strategic exclusion does also create unique points of intra-, as well as extra-ethnic conflicts, confusion, and contradictions. The great redemptive part of Malakpa's book, in the end, is its brutal, even if sometimes well-worn, truths.

OBI NWAKANMA
University of Central Florida
USA

Daria Tunca (ed.). *Conversations with Chimamanda Ngozi Adichie.*

Jackson, MS: University Press of Mississippi, 200 pp. $25.00
ISBN 978496829276, paperback

Nigerian novelist Chimamanda Ngozi Adichie reviewed President Barack Obama's memoir, *A Promised Land*, for *The New York Times*. Always a vocal fan of the Obamas (though perhaps more so Michelle than Barack), the review is an admiring one. She praises his prose as 'gorgeous' and 'always pleasurable to read' while discussing with scepticism Obama's 'savage self-questioning', wondering whether such questioning is itself a defence mechanism developed by the former American president to deflect criticism by anticipating it. Her book review carefully adheres to the form. For example, after detailing Obama's tendency to narrate every political decision with 'a watery considering of so many sides that resulted in no sides at all', she backs away from an extended critique of this facet of Obama's governing style, beyond its quality as a narrative device. Adichie in her role as book critic, views Obama's sins as stylistic, avoiding discussion of specific policy. However, her review ends with a somewhat broader meditation on the grace and admiration we afford public intellectuals like Obama (and perhaps Adichie herself), writing,

> it is now normal to preface any praise of a public figure with the word 'flawed,' but who isn't flawed? As a convention it feels like an ungracious hedge, a churlish reluctance to commend the powerful or famous no matter how well deserved. (*New York Times* 12 November 2020)

Having recently finished reading *Conversations with Chimamanda Ngozi Adichie*, I could not help but read these questions as being about Adichie's own emerging legacy as well. In many ways, this collection of 'conversations' feels entirely familiar to a reader versed in the general contours of the last fifteen years of African literature on the global stage. Adichie repeatedly grapples with the 'Image of Africa' in the West, the question of including Igbo words in her writing, and the primacy of The Caine Prize for African Writing. However, reading these interviews chronologically, a portrait of Adichie that is both inconsistent and self-consciously constructed emerges. Though she continually rejects the label 'celebrity' in favour of 'public person', (120) in many ways the interviews collected in this volume tell the story of a woman whose ambivalence towards the notion of celebrity is increasingly complicated by her own inclusion in that category. In her own words, 'I used to automatically think that there was virtue in the non-famous person and vice in the famous person' (128). She has clearly revised this assumption.

The book, comprising sixteen interviews conducted between 2003 and 2018, is unlikely to be a comfortable read for anyone, regardless of their level of appreciation for Adichie and her work. More conservative readers have always blanched at her rejection of gender roles and her participation in mainstream feminism. But those on the Left will also find her flawed. Her support for the Obamas and Hillary Clinton is firm and abiding even when faced with questions about Obama-era foreign policies that have harmed Africans. She is unapologetic for the various social media-driven controversies in which she has found herself, including the assertion that, for trans women, growing up without a menstrual period is somehow a privilege. She does not want to talk about Beyoncé. Her view of Nigeria's future is increasingly bleak.

As the journalists from *New African Woman* write in the preamble to their 2017 interview with the writer, '[t]his is Adichie – everyone wants a piece of her' (157). Beyond the clamour for her attention, photos, and autographs, the newly published volume, which is the first collection of interviews with the writer, demonstrates a reading public that increasingly wants to project their own agendas onto Adichie. In turn Adichie emerges as fiercely resistant to any efforts to be made to suit a narrative created by others.

Indeed Daria Tunca, the collection's editor, is transparent about the narratives that she herself has constructed for readers through her curatorial decisions. In her introduction, Tunca makes clear that the most compelling narrative that she sees emerging from the fifteen years of Adichie interviews is about a loss of nuance. As she says to Adichie in her own wide-ranging 2018 interview that concludes the volume, 'people have tended to move away from [Adichie's] work to focus more on the politics … the range of topics [she is] asked about has shrunk, and … the discussion often comes back to feminism' (183). Yet, while Adichie responds to this suggestion with agreement and concern, she explicitly rejects some of Tunca's other narrative impositions. When Tunca offers Adichie the theory laid out in her introduction, that this shift indicates that interviewers have moved away from treating her as a 'Native Informant', who can provide historical and sociological explanations of Nigeria, and towards treating her as a cultural expert, or even therapist, Adichie rejects Tunca's conflation of scholarly and journalistic interviewers with Q & A audiences.

Adichie's scepticism towards literary critics and scholars is well documented. In fact, at times the feeling seems mutual. Since about the time of *Americanah*'s release in 2013, there has been a troubling contradiction in African literature circles with regards to Adichie. While the writer's brand has found significant purchase with popular audiences in Nigeria as well as in the United Kingdom, the United States, and much of the world (*Americanah* has been licensed for publication in twenty-nine languages), scholars seem to puzzle over how to reconcile her impact in the African literary marketplace with her increasing celebrity. While some of this backlash often seems to be obviously gendered and in response to her role as style icon and beauty brand ambassador, recent African Literature Association annual conferences (of which Adichie was a keynote speaker in 2018) have seen an increasing animosity from a few high-profile African literature scholars towards her work. In particular, *Americanah* is often dismissed as her weakest book, too concerned with attracting an American audience over literary integrity. However, in a few of the interviews included in this volume, Adichie asserts that she resisted editorial input meant to shape the novel for American tastes. While she was told to make the work more subtle, remove Ifemelu's blogs, and avoid the explicit polemics of the work, she refused, and thought the novel might be a critical and commercial failure as a result.

In the context of her storied resistance to academics' interpretations of and impositions on to her work, it is difficult not to read some defensiveness in this story, which is repeated across multiple interviews. In 2004, she explained that she did not study English in university because 'I wanted to learn more than what to read into what I read' (7). And in 2018, she objects to a line of questioning from Tunca with the retort, '[t]he problem with academics is that you people don't want to see life as it is' (194). Thus, her counter-narrative of *Americanah* as a book she wrote without concession to the American critical gaze, takes on additional significance as yet another way for Adichie to break free of the academics' grasp.

To Tunca's credit, the volume omits the most notoriously controversial interviews without obscuring their significant impact on Adichie's public persona and reputation. Notably, Adichie's 2013 interview with Aaron Bady for the *Boston Review*, which was the source of great Twitter controversy due in part to her reference to one of the emerging Nigerian writers who had attended her Farafina writing workshop as 'one of her boys', is absent. However, readers of this collection who somehow missed the uproar the first time around may find themselves searching online to fill in the few blanks left by her lengthy discussion of the incident with Mazi Nwonwu from 2015. Nwonwu, himself a Farafina alumnus, is responsible for the longest and perhaps most revealing interview included in the volume, largely because he prompts Adichie to address these controversies head-on, and the easy intimacy between interviewer and subject seems to allow Adichie to be both honest and strikingly specific in her assessment of the situation.

As with many good stories, minor characters like Nwonwu populate this volume, revealing the protagonist's depth and sometimes stealing the spotlight for a short time. Adichie often references the late Binyavanga Wainaina, sometimes by his full name or just his first, sometimes only as her clearly beloved 'friend from Kenya' (172–173). The frequency with which she brings up his name suggests the tender affection between them. Another often-mentioned figure in her life is her father, also recently passed, and a towering figure in her world.

Another character, Professor Wale Adebanwi, is the only interviewer to appear twice in the collection. He interviews Adichie in 2004 and again in 2007, and these two conversations stand out among the earlier interviews as much for Adebanwi's style of questioning as for Adichie's responses. His interviews were conducted over email, and his questions are often long and detailed as a result, offering his personal interpretations of her work or meditations on Nigerian politics. However, while Adichie's responses sometimes playfully comment on Adebanwi's verbosity or more regressive gender politics, the reader gets a sense of eavesdropping on a familial conversation across a Sunday dinner table. A younger generation rolls its eyes at the anachronistic conservatism of the older; the older scoffs at the naïvety and capriciousness of youth. However, this serves to foreshadow another, more distressing narrative that emerges from this collection. Namely, Adichie's increasing alienation from Nigeria's ongoing story.

In the final interview, which was conducted especially for the book, Tunca asks Adichie about another interview controversy in which she has been embroiled: Adichie, talking to Hillary Clinton in the spring of 2018. Adichie had, cheerfully but pointedly, questioned why the first word in the former US Secretary of State's Twitter bio was 'wife', but former president Clinton's was not 'husband.' Clinton herself responded to the question good-humouredly, going so far as to change the bio; however, many Nigerians reacted with extreme anger and disapproval of Adichie's question. Adichie explained that many people, some of whom she had previously considered progressive, were outraged at her 'disrespect' for the institution of marriage. Even her family became targets of outrage and venom in response to the brief exchange. This experience seems to have deeply shaken the writer, disrupting her relationship to her home country.

It is tempting for scholars of various political inclinations to find this episode frustrating on multiple levels. Indeed, is it not superficial in itself for Adichie to worry about Hillary Clinton's Twitter bio when after all, this is a woman who has participated in imperialist foreign policies that have harmed the most vulnerable people in the world – people for whom Adichie is, for right or for wrong, often purported to speak. The whole episode risks seeming silly; however, when Adichie laments,

> I've always been a good daughter of Nigeria; I've always deeply loved Nigeria; I've always felt that Nigeria would work – and not just as an Igbo person, as a Nigerian person. I've always believed that we have problems, but that we can make it happen. After that, I really started to question it. I'm still questioning it (185)

It is impossible not to reevaluate the initial dismissal of the event. Can something that causes this magnitude of emotions truly be called superficial? Adichie herself would almost certainly reject the premise of such a question. Indeed, as this collection of *Conversations* demonstrates, Adichie is a figure who continues to resist political puritanism and simple narratives. As she reminds us, 'the world doesn't always align with your ideology' (167).

KATE HARLIN
Eureka College
Illinois

Ernest Emenyonu. *The Literary History of the Igbo Novel: African Literature in African Languages.*

London: Routledge, 2020, 150pp. $160.00
ISBN 9780367369613, hardback

Beyond communication, the Igbo language remains the primary tool of access and the bedrock of Igbo culture, Igbo literature, complex knowledge systems and identity. In 2006, the United Nations predicted the death of some Indigenous languages (the Igbo language included) in 50 years due to a decline in usage. Confronted with the muted treatment of the Igbo novel and loss of Igbo language, Ernest Emenyonu in *The Literary History of the Igbo Novel: African Literature in African Languages*, intervenes in this lacuna by examining the evolution, features, themes and the genealogy of the Igbo novel from 1857 to 2015. Significantly, the author's interest in the historical environment of the colonial era, the influence of Igbo worldview and knowledge systems first encoded in oral traditions and later transferred to the written form contributes to Igbo language studies, use and preservation. By tracing Igbo literary origins to diverse oral performances, the author anchors written Igbo literature on oral literature as an entity that existed before Western ideals of writing and Igbo orthography.

The author opens with a concise definition of the Igbo novel as a novel written in the Igbo language. Aside from prioritizing the language of expression, Emenyonu clarifies that such works 'organically grew out of Igbo oral tradition, depicting wholly and entirely Igbo worldview narrated in the skilled manner of Igbo orators' (3). Having encountered more than 120 Igbo novels, 50 plays, and over 24 scores of Igbo short stories collections, numerous memoirs, speeches and songs in Igbo Indigenous language in the course of the research, the author notes that it is imperative to study the evolution of the Igbo novel to probe its unique aesthetics and lineages. With the lingering threat of the loss of the Igbo language, a study of this magnitude opens up necessary conversations about African Indigenous language preservation and resurgence. And across each of its well-researched chapters, what comes through is the power of Igbo literature to sustain the people's cultural ethos and rekindle waning interest in Indigenous languages, worldviews, knowledge systems, and art forms.

Beyond celebrating the ground-breaking endeavours of Igbo novelists such as Tony Uchenna Ubesie, J.U.T. Nzeako, Chinedum Ofomatas and others, the author delves into their unique character portrayal, language sensitivity, and style. Of note is the chapter on the female voice and the rebuttal of patriarchy in the work of a female Igbo novelist – Julie Onwuchekwa. Through a critical analysis of her *Chinaagọrọm* (Ibadan: Evans Brothers, 1983), the author highlights the pivotal contribution of the novelist to the development of the genre and the shaping of contemporary narrative. In a novel that emphasizes the importance of female education as instrumental to female empowerment, Emenyonu notes that 'beyond empowerment, Onwuchekwa advocates female education as a right for all women folk, whether married or not' (114). Onwuchekwa's feminist inclination sets the novelist apart from her male peers and illustrates the immense contribution of female novelists to the growth of Indigenous literature.

The Literary History of the Igbo Novel addresses the evolution and distinctive peculiarities of the Igbo novel as well as its divergencies from others. According to the author, since 'the foundation of all Igbo literature is in its oral tradition' (14), contemporary forms like Igbo drama, novel and poetry are extensions of oral performances. And just as the oral performer relies on Igbo folktales, proverbs, and Indigenous stylistic devices to enrich tales and get his or her point across, so does the Igbo novel writer. Of note is that the first African language novel in West Africa – *Omenuko* – is written in Igbo language and was published in London in 1933 long before Nigeria's independence in 1960 and subsequent language policies. The Igbo novel, therefore, predates African stories written in European languages which commenced in the mid-twentieth century.

Due to the colonial impetus, Igbo orthography was instituted by missionaries eager to evangelize the natives. The author clarifies that 'the early missionaries were not interested in Igbo or other African languages for the sake of the languages' (39), but as the catalyst to propel Western ideologies and religion. Such efforts are recorded in phrasebooks, dictionaries, and vocabulary books by early missionaries like Rev. J.F. Schon and Rev. S.A. Crowther. The second phase of the Igbo language orthography included the collection of proverbs, riddles and oral narratives that were reconstructed with 'biblical information and religious instructions' (43) by the missionaries. But such efforts could only go so far. The author, therefore, commends the pioneering efforts of F.C. Ogbalu who through a grassroots association – Society for Promoting Igbo Language and Culture – took the language back to its users and laid the foundation for Igbo literary revival and language enrichment.

In recent years, the standardization of written Igbo and various Igbo dialects has been surrounded by controversies by Indigenous scholars such as Chinua Achebe, Donatus Nwoga, Nolue Emenanjo and others. Emenyonu contends that despite the significance of such disputes, the risk of stifling robust and free imaginative literary creativity is abundant. Informed by the knowledge that written Igbo literature is 'severely retarded in its subsequent development because of a … controversy over orthography' (83), the author posits that the mass production of Igbo written literature in any and every dialect could engender language resurgence. Said works could be thereafter be edited and published into standard written Igbo and other dialects where necessary to boost readership and language survival. According to the scholar, such efforts would undoubtedly mitigate language loss and inspire creative interest, output and curiosity. By so doing, Igbo literature and the language would benefit from coinages, aphorisms, and the introduction of words from several dialects.

The Literary History of the Igbo Novel opens a fresh frontier into the study of African Indigenous literature and Igbo literature specifically as it embraces Igbo language, folktales, novels, minstrelsy and interviews. Drawing on the peculiar history of the development of Igbo orthography and the movement from oral narratives to the written form, Emenyonu presents an expansive archive with excerpts, manuscripts and samples. Beyond the literary and stylistic analysis of various Igbo tales, there is a conscious effort on the part of the author to reproduce Igbo tales in the Igbo language thereby incorporating the Igbo language into a critical material written primarily in the English language. This laudable move recognizes the significance of Indigenous language to literature, language survival and decolonization. In research that spans continents (Europe and Africa) and decades with collated data and archival resources from the British Museum, and various libraries and archives, the author leaves no stone unturned in this seminal work. A timely and valuable contribution, this work will be beneficial to the field of African Indigenous literary studies, Igbo cultural studies and literature, Nigerian literature, and Igbo language studies.

KUFRE USANGA
University of Alberta
Canada

Jack Mapanje. *Greetings from Grandpa.*

Hexham, UK: Bloodaxe Books, 2016, 80pp. £9.95
ISBN 9781780373119, paperback

Greetings from Grandpa, Jack Mapanje's sixth volume of poetry is, in broad brush strokes, centred around politics, exile experience, and religion. The volume follows the lead of the poet's earlier volumes in their preoccupation with the political situation of Malawi but expands the theme to include both the political situation in Africa, especially in the Arab nations, and global politics and economy in Africa and the West. The collection also advances on the themes of exile common in the earlier volumes by the poet but juxtaposes home and exile repeatedly and focuses on experiences and perceptions of different people in diaspora. In *Greetings from Grandpa*, Mapanje affirms that home for him is both his family, nuclear and extended family, and Malawi and the village he left for the West without failing to state that there are some migrants who consider exile as home. The religious character of the collection proceeds from the focus in 'On His Divine Reprieve: A Confession', 'Thanksgiving', and 'Saved from Comrade Hippo's Grace' on the poet's affirmation that God is sovereign, and his acknowledgement of His intervention in human affairs, one of the poet's themes in *The Chattering Wagtails of Mikuyu Prison* and *Skipping Without Ropes*. The poet's faith-based perceptions of issues are accentuated by his deployment of religious motifs of confession and thanksgiving in the listed poems.

In *Greetings from Grandpa*, Mapanje explores different faces and phases of politics in Malawi in 'Our Anthology of Martyrs Thickens', 'Considering Our Golden Jubilees', 'The White Elephants of Home', 'Balamanja North Beach Revisited', 'Crossing Linthipe Bridge II', and 'Saved From Comrade Hippo's Grace', while he examines in 'When Egypt Went Up in Flame', 'Another Death So Mean: Libya', and 'The Arab Spring', the series of anti-government protests, uprising and armed rebellions across much of the Arab world in the early 2010s. International politics and West's hypocrisy is the poet's preoccupation in 'Watching Palestinians Being Butchered', and the same theme is alluded to in 'Another Death So Mean: Libya', and veiled in the concern, in 'The White Elephants of Home', with global economy and the disadvantaged position of the economy of the 'Third World' because of the collaboration of its leaders with global industrialists and despotism.

Although *Greetings from Grandpa* treats, in broad terms, politics, exile, and religion, each preoccupation is embroidered by a variety of themes. The exile theme is enriched, for instance, by issues such as the migrant's nostalgia for, thoughts of, or concerns about home, perception of where is recognized as home and, of course, survival in exile. Poems of titles that include 'home' as well as others like 'Greetings From Grandpa', 'Grandpa Travelling Again?', and 'Princess Alexandria Smiling (Luxembourg)' are on domestic, familial, and political themes, address question of home and define the concept in micro, macro, and global terms as the poet's family – nuclear and extended – his village, and finally the nation of Malawi. 'The Three People I Met in Diaspora', 'Surviving Freedom in Sunderland', 'The Carwash, Clifton Moor, York', and 'Load-shedding' are held together by their concern with survival in exile in their focus, among other things, on blacks in diaspora washing cars, driving commercial taxis, and wiping 'fat foreign bums' to survive against all odds.

The political concerns in 'Our Anthology of Martyrs Thickens', 'Considering Our Golden Jubilees', 'The White Elephants of Home', 'When Egypt Went Up in Flames', and 'Another Death So Mean: Libya' are woven around governance under African despots in Malawi, Egypt, Libya, and Uganda, the unsettling phenomenon of the political landscape of the continent littered with violence and death, and riddled with web of corruption of the political class, the looting of the state treasury, the appropriation and primitive accumulation of the wealth of the state by the political class, and their collaboration with the West, China, and other parts of Asia, impoverishment of the masses, state infrastructure decay, politics of division and tribalism, the dissatisfaction of the masses with, and protests against, bad governance, and the brute force these are met with by the state. The reader is however offered some respite by certain positive developments about the political landscape described in 'Somaliland Rebuilding, Hargeysa: An Outsider's View'. In that poem, Mapanje celebrates the choice by Somaliland to end civil wars with her southern neighbours and her determination to make Hargeysa, its capital city, a peaceful and well-fortified one.

Greetings from Grandpa also treats the role of the intellectual and the academy in African politics in its focus on the anxiety or failure of professionals to redeem the political situation respectively in 'When Egypt Went Up in Flames' and 'Considering Our Golden Jubilees'. In the latter poem, the poet laments the Pilate syndrome and the refusal of the 'comrades in litigation' to preside over a scandal involving one of their own while in the former poem he expresses anxiety over the enormous work of reconstruction that comrades will have to do after the protest in Egypt finally has ended.

'Kalikalanje of *Ostrich Forest*' is the longest poem in *Greetings from Grandpa*. A sequence, the poem combines the three subjects of politics, exile, and religion. While the sequence focuses dominantly on the conflict between Lisimu, a vulture-like-bird, and a family, Kanje, Kalikanje, and Kalikalanje, this is done against a political background. The question of politics in the piece is addressed in the juxtaposition of the regime of the crooked, malevolent, and heavy-handed paramount ruler of the villages and that of the libertarian, benevolent, and democratic one who succeeded him. The former's regime is full of ill will while the latter's is full of goodwill. The question of exile features in the sequence in, expectedly, the result of the evil regime and the flight of citizens in droves from the villages. The sequence appears a parable of Malawi of both Banda and post Banda years.

The preoccupation of 'Kalikalanje of *Ostrich Forest*' with religion is subtly introduced from the beginning of the sequence in the characterization of Kalikalanje as God's diminutive already fried one, and reinforced, most especially in 'Cantos XI', in two ways. The first is in the declaration of Kalikanje to her husband Kanje that some godly powers had set free-range eggs for her where she found them in *Ostrich Forest*. The second way is more definitive, and this is the appearance of the Fourth Figure during the encounter between Lisimu, the vulture-like bird, and Kanje's family of three. God is suggestively the Fourth Figure and His appearance affirms Mapanje's characteristic focus that God intervenes in the affairs of men when it is most dark and they are powerless. The appearance of the Fourth Voice compares with the experience of the three Hebrews in the Bible thrown into the fiery furnace.

Greetings from Grandpa is written in the vein of Mapanje's earlier poetry, but remains the most preoccupied with global politics and international experiences, religion, and exilic themes, mediated with thoughts about home. In the volume, Mapanje deploys humour, allusion, animal imagery, and oral traditional aesthetics generally associated with his poetry. *Greetings from Grandpa* underscores that Mapanje is still characteristically critical of political ills and satirical of foibles of institutions but the bitter and angry tone in the earlier collections by the poet has given way in the sixth volume to a mature and mellow voice of a grandpa. The description of the kleptomaniac political class as rodents – comrade rat and comrade hippo for instance

– is characteristically the humorous Mapanje, and so are the allusions to the Bible, history, and literature, and his deployment of animal imagery, folktales, and legends derived from Malawi oral lore and other traditions contiguous to it. The longest poem in *Greetings from Grandpa*, 'Kalikalanje of *Ostrich Forest*' is fertilized by a legend of the Yao-speaking African people of Malawi, Tanzania, and Mozambique, and the animal imagery and traditional symbols in the collection poetry – the chameleon, snakes, foxes, rats, and rodents – are derived from Malawian lore.

OLUFEMI DUNMADE
University of Illorin
Nigeria

Ada Uzoamaka Azodo and Akachi Adimora-Ezeigbo (eds). *Resident Alien and Other Stories: An Anthology of Immigrant Voices from Africa and the African Diaspora.*

Glassboro, NJ: Goldline and Jacobs, 2020, 169 pp. $7.99
ISBN 9781938598418, paperback

Resident Alien and Other Stories is an anthology of short stories by seven authors, six of who are women, and one a man. Each of them has contributed two stories to the collection. The authors' backgrounds reveal that they are experienced creative writers who have published variously, and some are reputable award winners. The wide-ranging preoccupations of the authors in the stories indicate an assortment of concerns in their different environments. The issues in the stories include migration, relocation, and the repercussions of both, the implications of transitional experiences on the different people involved, socio-cultural realities that impose specific demands on people, the difficulty in returning to the African motherland from the African Diaspora, and human behaviours, as well as people's reactions to the socio-cultural expectations of their environments. A dominant motif that underlies the stories is physical movement – migration.

Several of the stories highlight the disillusionment that migrant characters experience as their dreams and expectations get shattered when they confront the harsh realities of their new environments. 'Resident Alien', by Omofolabo Ajayi-Soyinka, is an epistolary that depicts the predicament of the narrator, a Nigerian, who moves to the United States of America with exotic dreams, which cause her to make promises to family and friends at home. She holds a Master's degree in Psychology and Social Welfare, has a twenty-year work experience from her home country, Nigeria, and had headed a department with thirty-five staff members. Unfortunately, all these do not facilitate her getting any of the prestigious jobs she desires and seeks to have in spite of her status as a legitimate permanent resident in the US. Her Green Card is of no consequence because she is regarded as alien, her Nigerian accent betrays her, and in two years of job-seeking, she looks seventy-five, rather than forty-five.

Ajayi-Soyinka's protagonist bears the frustration of being rejected in a foreign environment, just as does Teniola Oyenuga in Tomi Adeaga's 'The Expired Welcome'. Teniola gains admission to study International Relations at Johann Wolfgang Goethe University, Frankfurt am Main, Germany. In seven years, she acquires the German language, and obtains both Bachelor's and Master's degrees. She needs a job urgently, or to marry a German, to be able to stay on in Germany. All the application letters she sends out receive negative responses, mainly because she is black. She meets and falls in love with Fabian Himmelreich, who proposes to her, but his parents are firmly opposed to their only child marrying a black girl. Teniola bears the agony of double rejection that is based on acute racism. The paradoxical sentiment of the migrant character's dream being fulfilled by the relocation and hope is inconsistent with the disenchantment that emerges from the retrogressive and debilitating realities of the new environment.

The anti-climax that the migrant character is saddled with is often overwhelming and disorienting. That is what the eighty-year-old Mark Ude in Akachi Adimora-Ezeigbo's 'Dilemma of a Senior Citizen' feels in Edinburgh in the face of his incapacitation and near helplessness. Having divorced his wife, his children suggest that he should move into a nursing home because they are unwilling to be burdened by his needs for care and attention. Mark rejects the children's proposal, and decides to return to Nigeria and stay with his people. Despite the efforts of his brother, Onumba and wife, Mma to make him comfortable in a specially prepared Lagos apartment, he is unable to survive Lagos life – the gridlock, congestion, pollution, and more make him almost asphyxiated. Mark decides to move to the village, Umuokpa, where their eldest brother, Obioha, resides with his family. But Mark confronts a unique type of challenge in Umuokpa. There is no electricity because the rural electrification scheme had collapsed over a year before, resulting in most homes using power generating sets. Mark finds the noise and fumes unbearable. He also resents the constant invasion of his peace and privacy by extended family members who wake him with early morning knocks on his door just to greet him. Unable to cope with life in Umuokpa, Mark begins to query the wisdom in his return from Edinburgh. He decides to return there and concede to staying in the nursing home. Adimora-Ezeigbo subtly indicts Mark for having raised his two children – a son and a daughter – without exposing them to the culture or environment of Nigeria. Having divorced his wife, at age eighty, Mark has nobody to care for him, and cannot adjust to the Nigerian realities.

Mark Ude is a sharp contrast to Chisom in Akachi Adimora-Ezeigbo's 'Out of Sight', in which she demonstrates that being out of sight must not cause her to lose her integrity as a married lady, and a mother. A lecturer at the University of Lagos, Chisom is married to Dubem, an engineer, and together they have five-year-old twin sons. She obtains a six-month Commonwealth Fellowship to the Institute of Education, University of London. She is one of the five selected for the award from over one hundred applicants in her discipline. While in London, Chisom is almost seduced by Clifford (Cliff) from Ghana, whom she meets at a party, but she succeeds in escaping. Having regarded Cliff as a brother, she feels very disappointed at his attitude, and resolves to keep away from fellow Africans.

While Chisom struggles to sustain and uphold the credible values of the Nigerian culture, Tinuola's parents in Tomi Adeaga's 'Mirage' are carried away by the German cultural practices, which are inimical to the welfare and progress of a typical Nigerian child. On the one hand, Tinuola's mother, as a country club woman, has her group of friends with whom she spends a lot of time discussing frivolities about Nigerian social life; her father, on the other hand, is a homosexual. Her parents often disagree and fight. Tinuola has only the housemaid available to her. This unhealthy atmosphere affects the thirteen-year old Tinuola, and she is degenerating into a wreck, becomes violent and agitated, and performs poorly in school. The psychologist traces her attitude to parental neglect, and the trauma from knowing about her father's sexual orientation.

In Pede Hollist's 'A Life of Solitude', Khanu develops from being a child soldier in the Sierra Leonean Civil War, to being rescued by Peacekeeping forces, rehabilitated, made to acquire a skill, and then sent to America in a youth delegation to the United Nations General Assembly to give first hand information about life as a child soldier. During the war, Khanu once dared to tell his commander that he no longer wanted to be a soldier, and that earned him a sliced tongue, which healed him into silence, and compelled him into a world of solitude. At the end of the UN mission in America, Khanu defects from returning to Sierra Leone, slips into the shadowy world of America's undocumented immigrants, and a life of anonymity. His circumstances get him to marry Tamara, an American, who is twenty years older than him. After they have Bianca, Khanu and Tamara decide that Khanu should undergo a vasectomy yet, when Tamara later becomes ill, she is diagnosed as brain dead, and three-weeks pregnant. Khanu finds it difficult to reconcile Tamara's pregnancy with his sterile status. He establishes that she has been cheating on him, but his strategy of internalizing hurts helps him to bear the emotional trauma from the realization of Tamara's infidelity.

Naana Banyiwa Horne and H. Oby Okolocha focus on the importance that Africans attach to human life and death. Their stories portray the spiritual significance of human existence and relationships as well as the profound essence of death in African cosmology. In Horne's 'Miimi is Being Bad Again', set in Washington DC, and 'Reconnected Redefined', set in Miami, each of the Ghanaian ladies has a baby with a cleft lip and palate, that eventually dies. Significantly, in 'Reconnected Redefined', though DeJohn is an American married to Mansa and both live in Miami, he accepts the traditions of the Ghanaian community, and the beliefs of the people. He acknowledges the veracity of Chief Ajanu's notion that the new baby, Sunkwa, by Mansa and DeJohn is an *Abuku Kwasamba*, a baby born only to die prematurely, and get reborn to the same parents, and die again prematurely, and then keep repeating the agonizing cycle. Also Sunkwa has extra digits on the extremities, and suffers breathing difficulties as well as a defective thirteenth chromosome, which is a fatal condition. DeJohn and Mansa diligently follow Chief Ajunwa's guidance and recommendation that the baby be cremated to break the distressing cycle of birth/early death. They go to the African Temple, and the Chief performs some rituals and then empties Sunkwa's ashes into the Miami River, with the belief that the ashes will flow into the Atlantic Ocean and Sunkwa will eventually journey back to her Ghanaian roots.

Okolocha's 'The Londoner Goes Home' is an overt celebration of the life of Chiejina as his remains are brought home from London for burial. Conversely, the ceremony is a subtle condemnation of a titled Nigerian man, Chiejina, who spent about forty-five years in London, and at death, is brought home for burial. Sadly, Chiejina and his younger brother, Akunnia had not been on speaking terms for about twenty years. Akunnia feels unhappy that Chiejina had neither done anything for him or anybody else in the family, nor did he own a house in the homestead that would have been a source of pride to his family members. Additionally, Chiejina raised his son, Chude, to be a spectacle of shame to the family, as Chude, at almost sixty years old, is muscular and tattooed, and looking like a gangster, an image that is repugnant to the socio-cultural values and expectations of the Onitsha people.

Prominently, the motif of travel, physical movement and peregrination in the stories in *Resident Alien and Other Stories* offers a platform for the authors to narrate the challenges, transformations, setbacks, discriminatory attitudes, and disillusionment that the typical African migrant experiences. The stories open up vistas of configurations and expressions that bear marks of Africa and the African Diaspora. These are evident in the incidents that generate applause, admiration, condemnation, hope, and more, which are manifest in the stories.

Even though many of the stories have their complications resolved for a justifiable ending, this review focuses on the thrust of what the migrant character faces on leaving his/her African home for the new environment, from the challenge of maladjustment, through inability to meet the requirements for a decent employment, to racial discrimination, all of which combine to make the migrant a potential victim of the new system, and a failure. On the other hand, the African migrant to the African Diaspora is challenged by the realities of the African environment that make him/her unable to fit back into the African environment when the need for that arises and, even in death, the family members indict and disparage the corpse. The complexity of the portrayals in the stories is a result of the diverse experiences and backgrounds of the authors. The stories find relevance in the fact that emigration activities will continue to be undertaken by people all over the world, including Africans, for diverse reasons – migration or immigration – and it is imperative to note from these stories the implications and repercussions of the process. The stories constitute an oblique caution to African migrants to uphold the credible African values while in the new environments, and ensure to transmit those to their children for the sustenance of Africanness in the African Diaspora. This Afrocentrism is critically urgent at this time when African values are fast being eroded on the motherland and in the Diaspora by diverse foreign influences. The authors in this collection are obviously performing their fundamental obligation to Africa of advocating the perpetuation of African ideals and cultural values.

INIOBONG UKO
University of Uyo
Nigeria